MATHEMATICS
FOR THE
LIBERAL ARTS

MATHEMATICS
FOR THE
LIBERAL ARTS

JASON I. BROWN
DALHOUSIE UNIVERSITY
HALIFAX, CANADA

CRC Press
Taylor & Francis Group
Boca Raton London New York

CRC Press is an imprint of the
Taylor & Francis Group, an **informa** business

A CHAPMAN & HALL BOOK

CRC Press
Taylor & Francis Group
6000 Broken Sound Parkway NW, Suite 300
Boca Raton, FL 33487-2742

Printed on acid-free paper
Version Date: 20140930

International Standard Book Number-13: 978-1-4665-9336-7 (Pack - Book and Ebook)

Visit the Taylor & Francis Web site at
http://www.taylorandfrancis.com

and the CRC Press Web site at
http://www.crcpress.com

Printed on elemental chlorine free (ECF)
recycled paper containing 30% Post-Consumer Waste
Printed and bound in the USA

Dedicated to my girlfriend (and wife),

Sondra.

Contents

List of Figures

List of Tables

Preface

Mathematics for the Liberal Arts grew out of research that I carried out in 2004, applying Fourier transforms and logical reasoning to figure out how The Beatles played that wonderful opening chord of *A Hard Day's Night*. The interest in the general public garnered by the discovery was quite overwhelming—who knew so many people, many who hadn't seen mathematics since leaving high school, would find the story so fascinating? The Beatles were an easy entrance way—I may have to explain to some people why math is important, but I *never* have to explain why The Beatles are important! Further work of mine on math and music seemed to resonate with many, and I started work on a book, *Our Days Are Numbered: How Mathematics Orders Our Lives*, that illustrates the role mathematics plays in everyone's daily life. I also authored a monthly column in a newspaper on the same topic, and there was a lot of interest from the general public.

And that got me thinking about the math for liberal arts courses. Having looked over a number of texts on the subject, I felt that the textbook I wanted to write still hadn't been written, and I set to task on it. I wanted to create a text that focused on mathematics that undergraduates who aren't math majors (and indeed likely not science majors) could take away from the course and continue to use even outside university and college, in their professional and personal lives. So the topics I chose to cover stress this approach, ranging from conversion factors (a simple yet vital tool in life, I think), statistics, visualization, money and risk, to softer fields such as games, art, music and humor. The methods discussed aren't meant to be learned and shaken out of the head right after the final exam—I intend that students keep these skills, by practicing and applying them in their lives outside of the classroom. There is a movement to have mathematics for liberal arts courses meet various quantitative reasoning requirements, and this text was also written with this in mind. Other goals of the text are to develop a logical approach to data and reasoning that students might come across in life and through the media and to continue to be life-longer learners well beyond the confines of the course. For the most part, the chapters are self-contained, so the instructor can choose topics as desired.

In particular, I feel that non-math majors need good reasons for keeping the math rather than throwing it away. The best argument is to give them surprising ways that math can actually make them more productive, but also more creative. Many people may think that math and creativity are at opposite ends of the academic spectrum, but any mathematician—those who really know—will say otherwise. In fact, many mathematicians would consider their research an art rather than a science! Being a songwriter and guitarist as well as a mathematician, I know what an artist wants most is a **good new idea**. And adding a little math, taking a mathematical approach, can add just the innovative creative spark needed.

I have just a few quick notes about the setup of the book. Each chapter opens with a relevant story, one that relates to the content of the chapter, and one that I hope captures your interest. I try to build the suspense by drawing the story to a conclusion at the end of the chapter. I feel that too many math textbooks are only used as resources, only cracked open when a student needs to look something up. But I think that stories, real-life stories, are an excellent motivational tool for working through the math, and they show students exactly how mathematics can play a role in their lives. Besides, many of the students taking the courses are excellent readers, and why shouldn't the text play to their strengths as well? Marginal material should be read through— often there are historical background and mathematical insights you won't want to miss. There are white boxes with red outlines to indicate mathematical formulas or results that you'll want to refer to. And, the light red boxes are what I call "Life Lessons"—these are principles, derived from the mathematical approach, that you might want to consider and put into practice in your life outside of this course. Finally, I have a companion website, `www.jasonibrown.com/mla`, that has a number of additional examples (even musical ones, that you can listen to) along with some one-of-a-kind software programs I have written to support the material presented on art and music.

Acknowledgments

I'd like to thank a number of people. First, the acquiring editor at CRC Press, Sunil Nair, who understood my vision for the textbook. As well, I'd like to thank the other great folk at CRC, including Jessica Vakili, Marcus Fontaine, Robin Lloyd-Starkes, Amber Donley, Sharon Watson and Sarah Gelson. I had wonderful conversations with instructors across the U.S.—Joe Kudrle, Sheila O'Leary Weaver, Jenn Nordstrom, Krisan Geary and Gary Olson—and I so appreciated the constructive criticism from those who read drafts of the text—Liz McMahon, Gary Gordon, Jennifer Powers, Nancy Ann Neudauer, Keith Taylor and Karla Karsten. And of course, I'd like to thank my two sons, Shael and Zane, and especially my wife, Sondra, for all their support along the way.

I hope you enjoy the book, and think, think, think about how to apply the ideas and approaches outside the course to your lives. You'll be surprised just how far a little math can take you.

Jason I. Brown
Dalhousie University, Halifax, Canada

Introduction

YOU WERE ONCE A GREAT MATHEMATICIAN. A natural born mathematician. All babies are. We think of newborns as being a blank slate, but they arrive with some special skills that are built into their DNA. The most fundamental of these are the mathematical skills needed for learning.

Think about it. You go from no knowledge of the real world to, within a few years, a deep understanding of sounds, words, grammar, sight, objects, emotions, reasoning, and so on. You learn

- to categorize objects into types and pick out prototypes for each class, a process that depends essentially on forming equivalence relations (see Chapter 1),

- to infer properties of objects and consequences of actions, using experimental techniques and statistics (see Chapter 2),

- to make decisions based on desires (see Chapter 5),

- to strategize in order to win at games (see Chapter 5 as well), and

- to formulate abstract theories and representations of the real world, and think at higher levels about your own and others' thoughts (see Chapter 8).

You may have lost some appreciation for the mathematics you found essential (and loved!) as an infant and child—for many people that does indeed happen. Often this takes place at the hands of a poor mathematics teacher along the way. But that needn't be the case. You can get back to that inner child, the one who used mathematics seamlessly throughout the day. This book is about just that. You'll find that with a little more math, your daily lives can be more productive, more understandable and more creative. It's all up to you.

Part I

The Math in Your Life

IT'S HERE, THERE AND EVERYWHERE. It's hiding behind the nooks and crannies. It's there when you get up, it's there when you eat breakfast and start your day. It's there at work, and then again there when you relax at night. You can't avoid it. Mathematics surrounds you.

Now you can choose to ignore the math—that's up to you. But recognizing and embracing even a little of the mathematics that fills your days and nights can make life more understandable, more interesting. And it can make you—believe it or not—more creative. Just a little bit, and you can appear more suave in restaurants. You can be healthier and safer. and you can make more sense out of the data you read about. And all of that is good, good news.

Health, Safety and Mathematics

Air Canada Flight 143 was about halfway through its flight from Ottawa, Ontario, to Edmonton, Alberta, a distance of approximately 1,800 miles. It was a clear, beautiful day, and nothing was amiss or out of the ordinary to Captain Bob Pearson or his First Officer Maurice Quintal, or to the 61 passengers on board. Everyone was looking forward to touching down in Edmonton.

The plane was flying at an altitude of 41 000 feet over the small town of Red Lake, Ontario, north of Minnesota, and close to the border between Ontario and Manitoba, two central provinces in Canada. The calm was broken by a warning alarm that went off in the cockpit, signaling that fuel pressure for the left engine was dropping. The crew could only imagine that a fuel gauge malfunction was at fault, so they turned off the alarm. The on-board computer confirmed that there was enough fuel to complete the flight. Not a big problem. But then the warning alarm went off for the right engine, and a loud noise indicated that both engines were out of fuel, leaving the plane not only with no power, but with the instrument panels in the dark as well. The pilots could still use their radio, and contacted the control tower in Winnipeg. That city was too far away, but there was a small tarmac at an old air force base in Gimli, Manitoba. That would be their only chance of survival.

Earlier in the day, the plane's fuel gauges were indeed not working, and the crew had to manually calculate how much fuel the plane would need for the flight, using "drip" readings that would tell them the depth of the fuel in the tanks, and hence the volume in liters. The pilot and copilot hadn't been trained to do the calculation (on other jets, there were often flight engineers to carry out such tasks), but it seemed straightforward. The specific gravity of the fuel was written down by the maintenance refuelers as 1.77 pounds per liter, along with an estimate of the number of liters in the tank; Pearson and Quintal ordered a specific additional amount of fuel. Certainly, the fuel gauges should be fixed, but that could wait until after the flight. With plenty of fuel on board, Flight 143 was ready to go.

But back to the situation at hand. The pilots were in a predicament they had never envisioned—flying a Boeing 767 with no fuel and no power. A giant metal glider in the sky, descending down, down to earth. Who was to know that the catastrophe in the making was due to a small, albeit, crucial mistake? The lives of almost 70 people were in the balance, and all that stood between them and a fiery demise was the skill of the crew. And all because of a mathematical error . . .

You've been exposed to the notion of a number since you were a toddler. Counting on your fingers, counting with numbers, and then measuring with fractions and decimals. And of course, along the way, you learned to construct other numbers from the ones at hand—adding numbers, subtracting numbers, multiplying numbers, dividing numbers. Then you found how convenient it was to have general rules for building new numbers from old—functions. The level of abstraction grew, but at the heart of the matter were some very common life situations where all of the concepts were invaluable, and sometimes critical. This chapter will revisit some basic notions and show how a deeper understanding of numbers and functions can make a very big difference in your life.

1.1   *Found in Translation*

MATHEMATICS IS A LANGUAGE UNTO ITSELF. It has its own terminology, its own syntax, its own rules, its own written notation, its own expressions. And like any other language, you can translate everyday concepts into the language of mathematics, and conversely, translate mathematical language into everyday concepts.

In order to properly figure out the arithmetic that surrounds you, you need to be able to translate phrases in spoken language into the appropriate mathematics. There are some keywords in everyday speech that translate into arithmetic, and these are worth memorizing. (Don't worry, there are only a few of these. I once had a math professor who told me he went into mathematics because he had a bad memory!)

What's the Percent in That?

First, recall that percents correspond to decimals, after you move the decimal place two places to the left. So, for example, 72% as a decimal is 0.72 (since 72%, which is 72.% with a decimal point added, is 0.72 when you move the decimal two places to the left). As another example, 2.4% is 0.024 (when you move the decimal place two to the left, you need to add in an extra zero before the decimal point). On the other hand, 0.05 is the same as 5%, as you move the decimal point two the *right* when you go from a decimal to a percent.

One common translatable word is "of," which can have one of two translations into mathematics:

- The phrase "out of" converts to *division*. For example, "12 out of 20" corresponds to 12/20, which is 3/5, or 0.6 as a decimal.

- By itself, "of" translates to *multiplication*. For example, "half of 16" is 8, as $\frac{1}{2} \times 16 = 8$.

So combining our knowledge of percentages and the translation of "of", we see that, for example, 72% of 25 is $0.72 \times 25 = 18$. And, on the other hand, 41 out of 50 is $41/50 = 0.82$, or 82%. Arithmetic can be simple if you know how to translate.

Let's take our math into real life. You are out at a restaurant with a date and the waiter hands you the bill. There is a subtotal and a tax amount—the subtotal is $42.00 with tax of $3.36, for a total of $45.36. You know that you should leave a 15% tip on the meal cost, but not the tax. Do you know how to do the calculation?

What is the math involved? You want to leave a 15% tip on $42.00, that is, you need to calculate "15% of $42.00," which we know means finding 0.15×42.00. With a calculator you immediately find that the tip is $6.30. Voilà!

But what if you don't even want to take out your calculator, and you want to impress with your mental math skills (which of course will impress all dates)? There is a simple mathematical "trick" you can remember to do the calculation in your head. Notice that 15% is the same as 10% and 5%. So to calculate a

Percents are fractions as well, as fractions can be written as decimals—for example, 1/4 is 25%, as $1/4 = 0.25$.

Zeros can be added freely to numbers either before the first digit (provided the first digit is to the *left* of the decimal point, or to the right of the last digit, provided the last digit is *after* the decimal point. So, for example, 7.203 is the same as 007.203 and 7.203000. This comes in handy when converting from decimal to percentages and back.

Sometimes we represent multiplication with the × symbol, sometimes by a dot ·, and often just by putting two numbers side by side.

8

15% tip, you can break it up into 10% and 5% tips. Calculating a 10% tip is very easy—you simply move the decimal place one spot the left on the subtotal (as multiplying a number by 0.10 has this effect). So a 10% tip on $42.00, that is, 10% of $42.00, is $0.10 \times \$42.00 = \$4.20$. And the additional 5% tip is simply half of this, as 5% is half of 10%. So the 5% tip is $(1/2) \times \$4.20 = 2.10$, and you are probably pretty good at taking half of a number in your head. The total tip is the sum of the 10% and 5% tip which is $\$4.20 + \$2.10 = \$6.30$, and as promised, the calculation is simple enough to even do without any help!

The ability to take 10% of a number simply by moving the decimal place one spot to the left is very, very handy in many situations, beyond calculating tips. For example, if you are in a store that has a 20% off tag on an item costing $140.00, the dollar value of the discount does not require a calculator if you are "in the know." You simply take 10% of 140 by moving the decimal place of $140.00 one place to the left to get $14.00, and then double it to $2 \times \$14.00$ to get 20% of $140.00, that is, $28.00. The price of the item is then $\$140.00 - \$28.00 = \$112.00$.

Not so hard! And even if the numbers get jumbled in your mind, feel free to round a bit to get an estimate. So had the tag been 20% off an item costing $139.50, I could take 10% of it as $13.95, which I would then round to the nearest dollar, $14.00, and then double to get $28.00, as before. The answer wouldn't be exact, of course, but close enough (I'd just be off by a dime).

And had the discount been 27% on the original $140.00 item? Sans calculator, I could do one of two things:

- I could round the discount *up* to 30%, and then proceed by taking 10% of $140.00 to get $14.00, and then triple it in my head to get $3 \times \$14.00 = \$42.00$, or

- I could round the discount *down* to 25%, and then proceed by taking 10% of $140.00 to get $14.00, and then double it to get $2 \times \$14.00 = \$28.00$ and add in 5% more, or half of 10%, which is $7.00, to reach a discount of $35.00.

The actual discount is somewhere between $35.00 and $42.00. That gives you a pretty good idea of the discount.

Taking 1% of a number is just as easy—move the decimal *two* places to the left. So, for example, 1% of $7.00 is 0.07, that is, 7 cents, and 1% of $1,025.50 is $10.2550, that is, $10.26, when you round to the nearest cent.

A 20% **discount** means that you pay only 80% of the sticker price, a 30% discount corresponds to paying 70% of the sticker price, and so on.

Estimation is such a crucial life skill that you should employ it (and practice it) whenever and wherever you can, if only to check the reasonableness of the number calculated or presented.

What if you want to go for the gold and figure it out *exactly*, without a calculator? Even that is doable. Now 1% corresponds to moving the decimal place *two* places to the left, so 1% of 140.00 is $1.40. So we know that 25% of $140.00 is $35.00 and each extra 1% is $1.40, 27% of $140.00 is 25% + 1% + 1% of $140.00, that is,

$$\$35.00 + \$1.40 + \$1.40 = \$37.80.$$

A little harder, but still definitely something you can practice doing in your head.

Here is another scenario. What do you think about a $100 item that has been marked up 20% and then marked down 20%—surely nothing has changed, right? Wrong! Let's do the math to see what's going on. First, the 20% mark-up on the $100 item is $0.20 \times \$100 = \$20$. Pretty simple. So the item is now worth $120. But now marking it down by 20% reduces it by $0.20 \times \$120 = \$24$, so the price becomes $\$120 - \$24 = \$96$, less than what it started at! The lesson to learn is that it is very important what you are taking a percentage of, and as things change, the same percentage may amount to different amounts from earlier on.

Also, calculate the price of a $100 item that has been marked down 20% and then marked up 20%.

Going By the Rule

Now the act of multiplying by a fixed number—say 15%—can be thought of as just that—a mathematical act—but from another perspective, it can be envisioned as the application of a function to a number. A **function** is simply a rule that assigns to every object (in some collection) another object in another (possibly the same) collection. Often functions are given by formulas, but sometimes they can be given in other ways (such as tables or plots). So, for example, multiplication by 0.15 can be encapsulated by the function

One of the advantages of a mathematical mindset is that it shows that it is always helpful to look at an idea or process from more than one point of view.

$$f(x) = 0.15x.$$

The function's **domain**, that is, the permissible values that can replace the **variable** x here are all nonnegative real numbers, as representing costs of items (though a good case could be made for including negative real numbers in the domain as well, as you could take 15% of a loss as well). So, for example,

- 15% of 200 is $f(200) = 0.15(200) = 30$.

- 15% of 0 is $f(0) = 0.15(0) = 0$.

- If $f(x) = 450$, then $0.15x = 450$, so $x = 450/0.15 = 3000$, that is, if 15% of an unknown amount is 450, then the amount must be 3000.

- If y is an amount such that 15% of it was 125, that is, $f(y) = 0.15y = 125$, then 15% of twice that amount, $2y$, would be $f(2y) = 0.15(2y) = 2(0.15y) = 2f(y) = 2(125) = 250$.

Viewing the process as a function focuses more on the process rather than the numbers themselves. And we see that the notation of functions allows us to concisely write down the given information, and to solve for what is unknown.

While functions often produce from a single numbers other number, it need not be so simple, and herein lies one of the big strengths of mathematics. For example, I have found over the last few years that healthy eating (and keeping a healthy weight) has grown in importance for me. I have even joined a well-known weight management company to lose some pounds.

While details have changed slightly over the years, the idea behind this company and others like it is fairly simple. You are allotted a certain number of "units" or calorie goals during each day, and you use up units by eating, the number of units used depending on the food ingested. Now with a subscription, you are given a calorie goal converter, where you are asked, for each item you eat, to input the number of grams of carbohydrates, fat, protein and fiber, press an extra button, and presto—the number of units you have used up magically appears.

For example:

- From the "nutrition facts" on the label, 1/2 can of tuna has 0 grams (g) of carbohydrates, 1 g of fat, 16 g of protein and 0 g of fiber, and uses up 2 units.

- A 1/2 cup of mushroom soup has 9 g of carbs, 7 g of fat, 1 g of protein and 0 g of fiber, for 3 units.

- A 1/4 cup of pesto has 4 g of carbs, 23 g of fat, 4 g of protein and 0 g of fiber, for a whopping 7 units.

- A 2/3 cup serving of a certain breakfast cereal has 31 g of carbs, 8 g of fat, 5 g of protein and 4 g of fiber, for 6 units.

- 9 squares of dark chocolate with almonds has 25 g of carbs, 15 g of fat, and 3 g each of protein and fiber, for 7 units.

My daily allotment of units is 30, so I get more bang for my buck having tuna than chocolate—what a shame!

The expectation is that I carry my calculator around with me everywhere, to count units, but what if you are like me and often lose it or leave it behind? Are you out of luck? The answer is no, as, in fact, the units can be estimated by the formula

$$\text{cgunits} = \left\lceil \frac{19 \times \text{carbs}}{175} + \frac{9 \times \text{fat}}{35} + \frac{16 \times \text{protein}}{175} - \frac{2 \times \text{fiber}}{25} \right\rceil.$$

Here "cgunits" is a function, not of one variable, but of four—carbs, fat, protein and fiber (we could actually write the function more precisely as cgunits(carbs, fat, protein, fiber)) . The formula on the right side should look fairly ordinary, though it has fractions, except for the outside bars ⌈ and ⌉. These denote the fact that the answer should be *rounded up*, not to the closest whole number necessarily (I should mention that the units are indeed always whole numbers). So, for example, the breakfast cereal mentioned above, which has in a 2/3 cup serving 31 g of carbs, 8 g of fat, 5 g of protein and 4 g of fiber, for 6 units, would use up

Perhaps you are wondering how I came up with the formula? The formula was derived with *multiple linear regression*, and bears similar results to brand name weight management services, works extremely well in any instances I and others have applied it. You can run, but you cannot hide, from mathematics!

$$\begin{aligned} \text{cgunits}(31, 8, 5, 4) &= \left\lceil \frac{19 \times 31}{175} + \frac{9 \times 8}{35} + \frac{16 \times 5}{175} - \frac{2 \times 4}{25} \right\rceil \\ &= \lceil 5.56 \rceil \\ &= 6 \end{aligned}$$

while the 9 squares of dark chocolate with almonds uses up

$$\begin{aligned} \text{cgunits}(25, 15, 3, 3) &= \left\lceil \frac{19 \times 25}{175} + \frac{9 \times 15}{35} + \frac{16 \times 3}{175} - \frac{2 \times 3}{25} \right\rceil \\ &= \lceil 6.61 \rceil \\ &= 7. \end{aligned}$$

I should mention that the formula isn't quite right—if the number turns out to be negative (which will happen, for example, if it is all fiber—yuck!) then you should replace the unit value by 0; mathematically, you could rewrite the right-hand side of the function cgunits as

$$\max\left(\left\lceil \frac{19 \times \text{carbs}}{175} + \frac{9 \times \text{fat}}{35} + \frac{16 \times \text{protein}}{175} - \frac{2 \times \text{fiber}}{25} \right\rceil, 0\right).$$

So what are the advantages to keeping the function around rather than the calculator? There are many:

1. The formula is easy enough to memorize and remember (or to write down and keep in your wallet). That way, you can carry it with you wherever you go. It is impervious to dead batteries as well.

2. The formula tells you a lot how units are used up. For example, the number you multiply the fats by, $9/35 = 0.2571$, is more than twice that for carbohydrates, $19/175 = 0.1086$, showing you that each gram of fat uses up more than twice the number of units that a gram of carbs does, so given your daily budget of calorie goal units, it is better to choose something that increases the fat by a few grams than the carbs by the same amount. You also learn that fiber *deducts* units from what you eat, so increasing your fiber intake will serve you well by saving you units (and filling you up).

3. You could, if you wish, keep track of partial units by omitting the rounding up. I often do this, as I prefer to keep the units in my pocket rather than giving them up!

4. It also gives your brain a good workout.

Now the formula is a little complicated to work out without a calculator, so what can be done on the fly quickly? You can approximate some of the fractions. For example,

- $19/175$ is about 0.1 or $1/10$,

- $9/35$ is about 0.25, or $1/4$,

- $16/175$ is also about $1/10$, and

- 2/25 is about 1/10 as well.

So we could approximate our function's formula as

$$\text{cgunits} \approx \left\lceil \frac{\text{carbs}}{10} + \frac{\text{fat}}{4} + \frac{\text{protein}}{10} - \frac{\text{fiber}}{10} \right\rceil.$$

The symbol $\approx$ means "approximately equal to."

That's even easier to memorize and work with, especially since division by 10 just moves the decimal place one to the left, as we have seen. So, for example, the serving of breakfast cereal uses up about

$$\left\lceil \frac{25}{10} + \frac{15}{4} + \frac{3}{10} - \frac{7}{10} \right\rceil \approx \lceil 2.5 + 3.75 + 0.3 - 0.7 \rceil \approx \lceil 5.85 \rceil = 6,$$

which is a little off from the exact value of 7, but not by much.

> *Life Lesson:* Sometimes an easy, good approximation is worth more than an exhausting, precise answer.

Exercises

Exercise 1.1.1. Describe, in words, how to convert a decimal to a percentage.

Exercise 1.1.2. Describe, in words, how to convert a fraction to a percentage.

Exercise 1.1.3. Describe, in words, how to convert a percentage to a decimal.

Exercise 1.1.4. For each of the following, convert the percentage into a decimal value.
(a) 20%   (b) 1%   (c) 110%   (d) 0.01%

Exercise 1.1.5. For each of the following, convert the decimal value into a percentage.
(a) 0.35   (b) 0.035   (c) 2.1   (d) 100

Exercise 1.1.6. For each of the following, convert the fraction into a percentage.
(a) 1/10   (b) 2/3   (c) 3/4   (d) 9/100

Exercise 1.1.7. For each of the following, convert the wording into mathematics, but do not evaluate.
(a) a half of 20   (b) a third of 24   (c) a quarter of 15.2   (d) a tenth of 0.6

Exercise 1.1.8. Calculate the values for each part of Exercise 1.1.7.

Exercise 1.1.9. For each of the following, convert the wording into mathematics, but do not evaluate.
(a) 25% of 20   (b) 55% of 50   (c) 5.2% of 40   (d) 0.1% of 2000

Exercise 1.1.10. Calculate the values for each part of Exercise 1.1.9.

Exercise 1.1.11. Which would you prefer, to pay 75% of the sticker price, or to receive 20% off the sticker price?

Exercise 1.1.12. Which would you prefer, to pay 85% of the sticker price, or to receive 25% off the sticker price?

Exercise 1.1.13. Suppose you go out of state and enter a restaurant where the sales tax is 4%. Your bill is $55.00 before tax.
(a) What is the sales tax on your bill?
(b) What tip should you leave? (Round to the nearest nickel.)
(c) What is the total of your bill, tax and tip included?

Exercise 1.1.14. In some parts of Canada, the tax is 15%. Why does this make calculating the tip on a restaurant bill simple?

Exercise 1.1.15. Suppose the sales tax is 5.6%. Your bill is $121.00 before tax. Estimate your sales tax, but do not calculate it exactly.

Exercise 1.1.16. Suppose you are about to purchase a sofa with a sales price of $350.00. You notice a discount sticker of 20%. How much is the discount sticker worth? What is the discounted price of the sofa?

Exercise 1.1.17. If the sales tax for a state is 8%, what is the final price of the laptop, if the tax is calculated on the discounted price?

Exercise 1.1.18. If the sales tax for a state is 8%, what is the final price of the laptop, if the tax is calculated on full price and then the discount is applied?

Exercise 1.1.19. Suppose you tutor at a rate of $25 per hour. You give a discount of 25% on your hourly rate for friends. What do you charge them per hour?

Exercise 1.1.20. Suppose you tutor at a rate of $25 per hour. You add a surcharge of 25% on your hourly rate to people who annoy you. What do you charge them per hour?

Exercise 1.1.21. If the sales tax is 10.5%, write down the formula for a function that calculates the sales tax on an item with price x dollars.

Exercise 1.1.22. If the sales tax is 10.5%, write down the formula for a function that calculates the total cost of an item with price x dollars.

Exercise 1.1.23. If the sales tax is 8%, write down the formula for a function that calculates the total cost of an item with price x dollars.

Exercise 1.1.24. If the sales tax is 10%, and an item has a final cost of $63.25, what was the price of the item, before tax?

Exercise 1.1.25. If the sales tax is 11.25%, what happens to the tax if you triple the price of the item? Does this depend on the tax rate?

Exercise 1.1.26. If the function $f(x) = 0.027x$ calculates the sales tax on an item with ticket price x dollars, what is the sales tax rate?

Exercise 1.1.27. If the function $f(x) = 1.085x$ calculates the final price (including sales tax) on an item with ticket price x dollars, what is the sales tax rate?

Exercise 1.1.28. Suppose an item that costs x dollars is marked up $r\%$ and then marked down $r\%$. Can you say whether in the end it cost more, less or the same?

Exercise 1.1.29. Suppose an item that costs x dollars is marked up $r\%$ and then marked down $r\%$. Another item that also costs x dollars is marked *down* $r\%$ and then marked *up* $r\%$. How do the two items compare in cost? Why?

Exercise 1.1.30. You undoubtedly have heard many a sports figure say that he or she has given 110%, which is, of course, nonsense. But what if a player says that she gave 25% less than her maximum yesterday, but gave 120% today? Does this make sense? If so, what percentage of her best is she playing at today?

Exercise 1.1.31. The U.S. debt is approximately $\$17\,528\,000\,000\,000$, that is, about 17.528 *trillion* dollars. The population of the U.S. is approximately $316\,150\,000$. What is the amount of debt per person?

Exercise 1.1.32. From the calorie goals formula on page 11, if you have a choice between a certain number of grams of fat, carbohydrates and protein, which should you choose? Why?

Exercise 1.1.33. Consider the calorie goals function

$$\text{cgunits} = \left\lceil \frac{19 \times \text{carbs}}{175} + \frac{9 \times \text{fat}}{35} + \frac{16 \times \text{protein}}{175} - \frac{2 \times \text{fiber}}{25} \right\rceil.$$

(a) If the nutritional label tells you that a serving has 20 g of carbs, 12 g of fat, 11 g of protein and 5 g of fiber, how many units is a serving?

(b) How many units are 3 servings of the item described in part (a)?

Exercise 1.1.34. Using the function cgunits described in Exercise 1.1.33:

(a) If the nutritional label tells you that a serving has 15 g of carbs, 21 g of fat, 4 g of protein and 0 g of fiber, how many units is a serving?

(b) If you have 27 units left for the day, how many servings of the item in part (a) can you eat?

Exercise 1.1.35. Pick up some packaged items at the supermarket. Using their nutritional guides, calculate the calorie goal units for each item.

Exercise 1.1.36. Estimate (without using a calculator!) the units of the item in Exercise 1.1.33(a).

Exercise 1.1.37. Estimate (without using a calculator!) the units of the item in Exercise 1.1.34(a).

Exercise 1.1.38. Under a new weight management system, most fruits and vegetables (like apples, oranges, celery, carrots, etc.) are *free*—you don't have to keep track of their units. Why? Can you think of some fruits and vegetables for which that should not be the case?

Exercise 1.1.39. The older version of the calorie goal formula (pre-2010) is given by

$$\text{oldcgunits} = \text{round}\left(\frac{\text{calories}}{50} + \frac{\text{fat}}{12} - \frac{\min(\text{fiber}, 4)}{5} \right),$$

where "round" denotes rounding to the nearest integer, and "min" takes the smaller of the two values (again, fat and fiber are measured in grams).

(a) Compare the two formulas. What do you think are the advantages of the new calorie goal function over the old one?

(b) A serving of crackers has 90 calories, 3.5 g of fat, 13 g of carbohydrates, 1 of fiber and 2 g of protein. How many units is it under the old and new calorie goal formulas?

(c) A serving of tuna has 70 calories, 1 g of fat, 0 g of carbohydrates, 0 of fiber and 16 g of protein. How many units is it under the old and new calorie goal formulas?

1.2 The Essentials of Conversion

QUANTITIES, LIKE WORDS, SOMETIMES NEED TO BE CON-
VERTED. People in different locations or under different circum-
stances take measurements in different scales.

Success Is Proportional to the Work

Definition 1.2.1. We say that measurement (or quantity) y is
proportional to measurement (or quantity) x if there is some
fixed positive number C (called the **constant of proportionality**)
such that
$$y = Cx.$$

We also call C the **conversion factor** from x to y. When y is
proportional to x we sometimes write $y \propto x$. The important
fact is that if y is proportional to x, then multiplying one by a
factor multiplies the other by the same factor. So, for example,
doubling x doubles y, tripling x triples y, and halving x halves
y, and so on. Here are some examples of quantities that are
proportional:

Proportionality is such a useful property when it holds. How do we check if two real-life quantities are proportional? You gather data and analyze it. More about this in Chapter 3.

- distance in miles and feet

- distance in miles and kilometers

- length in feet and inches

- height in meters and feet

- area in square meters and square yards

- volume in liters and gallons

- weight in pounds and kilograms

- time in minutes and seconds

- time in hours and years

- dollars in U.S. and Canadian funds

- U.S. dollars and Euros

Proportionality is a **relation**—a way of relating two objects. Two quantities are either proportional or not. Notice that

- any quantity x is proportional to itself (as $x = 1x$ always holds),

  $x \propto x$

- if quantity y is proportional to quantity x, then x is proportional to y (since, if $y = Cx$ then $x = \frac{1}{C}y$, so the constant of proportionality changes from C to $\frac{1}{C}$, but is still a fixed positive number), and

  $y \propto x$ implies $x \propto y$

- if quantity z is proportional to quantity y and quantity y is proportional to quantity x, then quantity z is proportional to quantity x (since if $z = C_1 y$ and $y = C_2 x$, then $z = C_1(C_2 x) = (C_1 C_2)x$, and the constant of proportionality of z and x is the products of the two constants of proportionality).

  $z \propto y$ and $y \propto x$ implies $z \propto x$

These three properties—any object x relates to itself, if x relates to y then y relates to x, and if x relates to y and y relates to z then x relates to z—are known as **reflexivity, symmetry** and **transitivity**, respectively, and crop up in many areas of mathematics. From the second property, we see that the order makes no difference for proportionality, so we can merely say x and y are proportional, rather than say y is proportional to x or x is proportional to y. Whenever you have a relation on some objects that is reflexive, symmetric and transitive, it breaks up all of the objects into groups, where objects in the same group are related, and objects in different groups are completely unrelated. Proportionality does exactly this—the different measurements for length/distance are in one group, being proportional to one another, measurements for area in another, and so on.

Relations that are reflexive, symmetric and transitive are called **equivalence relations**, and they fill the landscape of much of mathematics. And the whole process of categorizing objects as an infant (or older) depends on defining appropriate equivalence relations.

You may have noticed that the formula $y = Cx$ reminds you of straight lines, and indeed this is a visual way to think about

18

proportionality—two quantities x and y are proportional if the plot of y versus x is a straight line *through the origin*: the formula $y = Cx$ tells you that the y-intercept is 0 and the slope is C, the constant of proportionality.

So different units for measuring things like length, distance, time, speed, area, volume and currency are always proportional. Are all units for measuring proportional? No—one that comes to mind is Fahrenheit (F) and Celsius (C) for temperature. For example, 32°F is 0°C, but doubling the temperature to 64°F does *not* correspond to a temperature of $2 \times 0°C = 0°C$ (64°F actually corresponds to approximately 17.7°C). The formula for converting from Fahrenheit to Celsius is

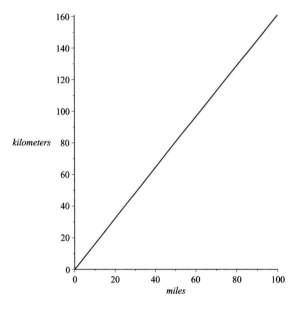

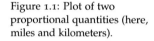

Figure 1.1: Plot of two proportional quantities (here, miles and kilometers).

$$\text{Celsius} = \frac{5}{9} \times (\text{Fahrenheit} - 32).$$

Expanding out the right side we see that

$$\text{Celsius} = \frac{5}{9} \times \text{Fahrenheit} - \frac{160}{9}.$$

This formula is indeed a straight line (with slope 5/9), but does not pass through the origin $(0,0)$, and that is why the Fahrenheit and Celsius temperatures are *not* proportional.

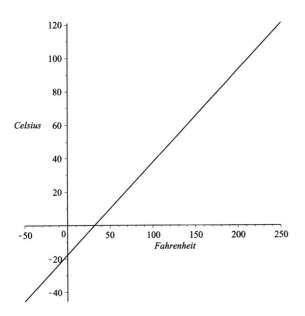

Figure 1.2: Temperatures in Celsius versus Fahrenheit. Note how the line does *not* go through the origin.

Many simple real-life applications of mathematics involves converting from one set of units to another. For example, if you take a road trip from the U.S. into Canada, you'll find that the distances change from miles to kilometers. Realizing that distance measurements are all proportional, you need to convert to use your odometer, and to do so, you need to know the constant of proportionality. The basic principle is the following:

> To find the conversion factor when going from measurement x to measurement y, form a fraction of equal amounts, with a quantity in units for y in the numerator, and an equal quantity in units for x in the denominator.

Some examples will drive home the process. Suppose you want to convert from miles to kilometers. You need to know that 1 mile is equal to 1.6 kilometers. Thus the conversion factor *from miles to kilometers* is

$$\frac{1.6 \text{ kilometers}}{1 \text{ miles}}.$$

Now, of course, the number in the fraction is simply 1.6, so isn't this the conversion factor? Yes and no. Yes, it is the right number, but no, if you just carry around the number you are prone to making some costly mistakes. The key thing to remember is:

There is a prototype of a bar, created in 1889 and held at the *Bureau International des Poids et Mesures* in Sèvres, France, and it has the "standard" length of 1 meter.

> When you work with conversion factors, **always** keep the units!

Leaving out the units, while it might seem convenient (especially if you have worked with all of the units many times before), is a recipe for disaster—more about this later! There is a whole branch of mathematics called **dimensional analysis**. When you add or subtract two quantities, their units must be the same, and the result is also in the same units. When you multiply two quantities, you multiply the units as well; when you divide one quantity by another, you divide the corresponding units. These facts are *critical* to working properly with units. You should <u>always</u> keep track of all units and cancel when appropriate. The guiding principle in dimensional analysis is:

Life Lesson: If the units of the final calculation aren't what's expected, you must have done the math wrong!

So suppose on our road trip we want to convert a distance of 352 kilometers (km) to miles (mi). There are two conversion factors available for us to use:

$$\frac{1.6 \text{ km}}{1 \text{ mi}} \quad \text{and} \quad \frac{1 \text{ mi}}{1.6 \text{ km}}.$$

Which do we use? Rather than trying to memorize[1] which one to use, we realize that as the quantities are proportional (distance in miles and distance in kilometers), we will get our answer by multiplying our known quantity, 352 kilometers, by one of the conversion factors: either

$$352 \text{ km} \times \frac{1.6 \text{ km}}{1 \text{ mi}}$$

[1] which we'll probably get wrong half the time!

or

$$352 \text{ km} \times \frac{1 \text{ mi}}{1.6 \text{ km}}.$$

In the first possibility, there is no canceling of units, and by multiplying out the numbers and units, we get

$$352 \times 1.6 \frac{\text{km}^2}{\text{mi}} = 563.2 \frac{\text{km}^2}{\text{mi}},$$

which can't be right, as the units are *not* miles. The second possibility gives, with unit canceling,

You multiply and divide units as you would numbers, so that, for example,

$$\text{km} \times \text{km} = \text{km}^2.$$

$$352 \text{ km} \times \frac{1 \text{ mi}}{1.6 \text{ km}} = \frac{352}{1.6} \text{ mi} = 220 \text{ mi},$$

a reasonable answer, units-wise, and indeed the correct one. You might be surprised at how many people, even knowing the conversion factors, pick the wrong one, simply because they ignore the units.

You may find it easier to remember that 50 miles is about 80 kilometers instead of 1 mile is equal to 1.6 kilometers (many cars have speedometers that show both miles per hour and kilometers per hour, and 50 miles per hour lines up with 80 kilometers per hour). Thus instead of the conversion factor

$$\frac{1 \text{ mi}}{1.6 \text{ km}}$$

we can use

$$\frac{50 \text{ mi}}{80 \text{ km}}.$$

The calculation looks different, but the result is the same:

$$
\begin{aligned}
352 \text{ km} &= 352 \text{ km} \times \frac{50 \text{ mi}}{80 \text{ km}} \\
&= \frac{352 \times 50}{80} \text{ mi} \\
&= 220 \text{ mi}.
\end{aligned}
$$

You could also use the conversion factor

$$\frac{100 \text{ mi}}{160 \text{ km}},$$

as 100 miles is about 160 kilometers. They amount to the same thing—all that is required is that you know some amount in both sets of units.

Some people remember that 100 kilometers is the same distance as 60 miles, but the resulting conversion factor is $\frac{60 \text{ mi}}{100 \text{ km}} = 0.6 \frac{\text{mi}}{\text{km}}$, which is different from $\frac{1 \text{ mi}}{1.6 \text{ km}} = 0.625 \frac{\text{mi}}{\text{km}}$. They are close, but the first is less accurate, and may make a significant difference for converting a large number of kilometers to miles.

Let's consider some other examples, first to having a healthy life. Suppose you are making dinner for nine guests, and your special soup recipe calls for 3 cups of vegetable stock and 2 teaspoons of salt, among other ingredients. The problem is, the recipe serves four, and you need to serve nine! What to do? The key thing is that the ingredients are obviously proportional to the number of servings—if you double the number of servings, you'll have to double the ingredients. So, for example, the 3 cups of vegetable stock is *per* 4 servings, so you get the conversion factor

$$\frac{3 \text{ c}}{4 \text{ srv}}$$

for converting from servings (svg) to cups (c) of vegetable stock. Thus for 9 servings, we require

$$9 \text{ svg} \times \frac{3 \text{ c}}{4 \text{ srv}} = \frac{27}{4} \text{ c} = 6.75 \text{ c}.$$

That is, we need six and three-quarter cups of vegetable stock for your excellent soup to serve nine. The same approach will tell you that you need

$$9 \text{ svg} \times \frac{2 \text{ tsp}}{4 \text{ srv}} = 4.5 \text{ tsp}$$

of salt, using a different conversion, relating teaspoons (tsp) of salt to servings. Going through all the ingredients in this way will ensure that you get the amounts just right!

Our next example involves something a bit more serious, drug dosing. Often (but not always, so please be careful!) drugs are administered according to the patient's weight. Suppose you are a camp counselor for a group of 15-year-old kids, and everyone has been outside enjoying the great outdoors. Unfortunately, one of them has been scratching like crazy because of mosquito bites. She weighs 124 pounds, and the packaging inside the antihistamine box says to administer 10 ml of the medication per 95 pounds of weight. This gives you a conversion factor of

$$\frac{10 \text{ ml}}{95 \text{ lb}}$$

between pounds of weight and milliliters of medication. So the

Here are some common abbreviations, inside the kitchen:

measurement	abbreviation
tsp or t	teaspoon
Tsp or T	tablespoon
c	cup
qt	quart
gal	gallon
ml or mL	milliliter
L	liter
oz	ounce
lb	pound
g	gram
kg	kilogram

correct dosage would be calculated as

$$124\ \cancel{\text{lb}} \times \frac{10\ \text{ml}}{95\ \cancel{\text{lb}}} = 13.05\ \text{ml}.$$

Now if you don't have a syringe with you, you can use a tea-spoon, as there are 5 ml in a teaspoon. This gives you another conversion factor, one to convert from milliliters to teaspoons, so choosing the right one to get the units to cancel properly, we find that we should give

$$13.05\ \cancel{\text{ml}} \times \frac{1\ \text{tsp}}{5\ \cancel{\text{ml}}} = 2.6\ \text{tsp}$$

of medication. Rounding down (it is safest to underdose than overdose!), we can give two and a half teaspoons of the antihistamine and wait for things to calm down for the camper.

Currency exchange rates vary over time, but at the time I am writing this paragraph, the currency exchange rate between Canadian dollars (CAD) and U.S. dollars (U.S.D) is given by

$$1\ \text{CAN} = 0.9048\ \text{U.S.D,}$$

that is, 1 Canadian dollar is equal to 0.9048 U.S. dollars. So we get the conversion factors

$$\frac{1\ \text{CAN}}{0.9048\ \text{U.S.D}} = 1.1052\ \frac{\text{CAN}}{\text{U.S.D}}$$

and

$$\frac{0.9048\ \text{U.S.D}}{1\ \text{CAN}} = 0.9048\ \frac{\text{CAN}}{\text{U.S.D}}.$$

Which one we apply depends on the right units canceling. For example, to convert 350 U.S.D to CAN, we calculate

$$350\ \cancel{\text{U.S.D}} \times \frac{1\ \text{CAN}}{0.9048\ \cancel{\text{U.S.D}}} = 386.83\ \text{CAN.}$$

Note how the units canceled appropriately to leave the answer in Canadian dollars. On the other hand, to convert 125 CAN to U.S.D, we proceed as follows:

$$125\ \cancel{\text{CAN}} \times \frac{0.9048\ \text{U.S.D}}{1\ \cancel{\text{CAN}}} = 113.10\ \text{U.S.D.}$$

Currency exchange rates fluctuate from day to day (and even minute to minute). Since 1950, the highest rate was on January 18, 2002, when 1 U.S. dollar was worth 1.6125 Canadian dollars, and the lowest was on November 6, 2007, when 1 U.S. dollar was only worth 0.9215 Canadian dollars.

You should always ask yourself in the end, *does the answer make sense*? Canadian dollars (as of the writing of this text!) are worth less than U.S. dollars, so that if we convert from U.S.D to CAN we expect more, and converting from CAN to U.S.D we expect less. The examples illustrate this.

As a final example, imagine you own a pizzeria and are pricing out your "fully dressed" round pizzas, that is pizzas with all of the toppings. It seems reasonable, *from a consumer's point of view*, to price the pizza out by the area of the pizza, as the area determines the amount of crust, cheese, sauce and toppings required (pizzas <u>aren't</u> priced this way in real life, as we'll soon see—you pay a premium for smaller pizzas). In fact, the amounts of each are proportional to the area of the pizza—if the pizza has twice the area, it will require twice the crust, cheese, sauce and toppings. A pizza that is twice the area should cost twice as much. So the price P of a pizza should be proportional to the area A of the pizza (which we'll measure in square inches, that is, in^2).

The area of a pizza is not so easy to measure directly. But typically, you measure a round pizza from one side to another, that is, you measure the **diameter** of the circular pizza. You probably remember that if you are given the radius r of a circle, the area A is given by

$$A = \pi r^2.$$

But the diameter D is twice the radius, that is, $D = 2r$, so

$$r = \frac{D}{2}.$$

Figure 1.3: A circle, its radius and diameter.

Substituting in to the formula for the area, we get

$$A = \pi \left(\frac{D}{2}\right)^2 = \frac{\pi}{4}D^2 = CD^2,$$

where C is the constant $\frac{\pi}{4}$. This shows that the area of a pizza is proportional to the *square* of the diameter, that is, $A \propto D^2$. So if you double the diameter D, you multiply the square of the diameter by 4, since $(2D)^2 = 4D^2$. That is, you *quadruple* the square of the diameter, and thus you quadruple the area of the pizza, and should quadruple the price of the pizza.

Suppose that you've set the price of an 8 inch pizza to $8.99. What should you charge for a 16 inch pizza? It has twice the diameter, and hence four times the square of the diameter. As the area is proportional to the square of the diameter, the area has quadrupled as well, that is, is multiplied by a factor of 4. We have stated that the price should be proportional to the area, so since the area has quadrupled, so should the price. Thus the price for our pizza should be

$$4 \times \$8.99 = \$35.96.$$

But let's work with the units. The area of an 8 inch pizza is $\pi \times (8 \text{ in})^2 = 64\pi \text{ in}^2$ (if you want, you can approximate π with 3.14 and get $64 \times 3.14 = 263.76 \text{ in}^2$, but I prefer to keep π hanging around, hoping that it will cancel out before I am forced to approximate it). As price and area are proportional, we can write out the conversion factors

$$\frac{8.99 \text{ dollars}}{64\pi \text{ in}^2} \quad \text{and} \quad \frac{64\pi \text{ in}^2}{8.99 \text{ dollars}}.$$

(We probably won't need both, but they are available.) Now a 16 inch pizza has area $\pi \times 16^2 = 256\pi \text{ in}^2$, so a pizza this size should cost

$$256\pi \text{ in}^2 \times \frac{8.99 \text{ dollars}}{64\pi \text{ in}^2} = \frac{256\pi \times 8.99}{64\pi} \text{ dollar} = 35.96 \text{ dollars}.$$

Students often ask me if they need to keep track of the units, and I say no—you also don't need to get the correct answer, either! The use of units may seem overkill, but they ensure that you use the right conversion factor, properly.

Similar Is as Similar Does

Note that while we related price to area for the pizzas, the area A was not immediately measurable—a different quantity, diameter D, was, and that is what we used. We saw that $A \propto D^2$, that is, the area was proportional to the *square* of the 1-dimensional diameter measurement. In fact, the area is proportional to the square of other 1-dimensional measurements, such as the radius or even the circumference of the pizza. And the proportionality of the area to the square of a 1-dimensional measurement is true even for non-round pizzas, as long as all of the pizzas have the same shape. This observation is so important that we'll point it out explicitly.

Definition 1.2.2. Two objects are **geometrically similar** if they have the same shape, that is, one is a scaled version of the other. For a group of geometrically similar objects, an easy to measure quantity is called a *characteristic dimension* for the objects.

Figure 1.4: Geometrically similar objects.

Circles are all geometrically similar, as are all squares, while rectangles are not (long rectangles have a different shape from squares, which are rectangles). A characteristic dimension for a circle is the radius (the diameter or circumference would do as well); for a square, the side (or the diagonal) would be a characteristic dimension. These characteristic dimensions are all of dimension 1, as they just have a length and no width. One of the key facts is that if we have a group of geometrically similar objects with a characteristic dimension, say x, of dimension 1, then

- an object's area is proportional to x^2, the *square* of the characteristic dimension, and

- an object's volume is proportional to x^3, the *cube* of the characteristic dimension.

So, for example, the area of circular pizzas (which are geometrically similar, as they all have the same shape) is proportional to the square of the diameter, a 1-dimensional characteristic dimension for circles. Had we decided to make instead square pizzas (which are geometrically similar), we would find that the area of the pizzas would still be proportional to the square of the length of a side, a 1-dimensional characteristic dimension for squares. And again, were we to double the characteristic dimension for square pizzas, say double the length of the sides, the areas would quadruple, as should the price.

It is interesting to actually see how pizzas are priced. At a local pizzeria, I find that

- a 10-inch pizza is $13.24,

- a 12-inch pizza is $16.89,

- a 14-inch pizza is $20.59, and

- a 16-inch pizza is $24.99.

If we use for our basis the smallest pizza, which has area $\pi \times (10 \text{ in})^2 = 100\pi \text{ in}^2$, then the appropriate conversion between price and area is

$$\frac{13.24 \text{ dollars}}{100\pi \text{ in}^2}.$$

Based on this conversion factor, the price of

- the 12 inch pizza should be

$$\pi \times 12^2 \text{ in}^2 \times \frac{13.24 \text{ dollars}}{100\pi \text{ in}^2} = 19.05 \text{ dollars},$$

- the 14 inch pizza should be

$$\pi \times 14^2 \text{ in}^2 \times \frac{13.24 \text{ dollars}}{100\pi \text{ in}^2} = 22.95 \text{ dollars},$$

and

- the 16 inch pizza should be

$$\pi \times 16^2 \text{ in}^2 \times \frac{13.24 \text{ dollars}}{100\pi \text{ in}^2} = 33.84 \text{ dollars}.$$

So we see that there are big savings to be had by ordering larger pizzas (and that indeed, the π's canceled out!).

On the other hand, if we base everything on the 16-inch pizza, the conversion factor is

$$\frac{24.99 \text{ dollars}}{256\pi \text{ in}^2},$$

and based on this conversion factor, the price of

- the 10-inch pizza should be

$$\pi \times 10^2 \text{ in}^2 \times \frac{24.99 \text{ dollars}}{256\pi \text{ in}^2} = 9.76 \text{ dollars},$$

- the 12-inch pizza should be

$$\pi \times 12^2 \,\text{in}^2 \times \frac{24.99 \text{ dollars}}{256\pi \,\text{in}^2} = 14.06 \text{ dollars},$$

and

- the 14-inch pizza should be

$$\pi \times 14^2 \,\text{in}^2 \times \frac{24.99 \text{ dollars}}{256\pi \,\text{in}^2} = 19.13 \text{ dollars}.$$

So we see from the store owner's point of view, the most profit is to be made on the smallest pizza.

Conversion factors may help choose a pizza, but can also be useful in making healthier choices. The proportionality that we've seen so far has sometimes involved quantities of the same type—for example, distance and currencies—and sometimes quantities of different types—such as price and area for pizzas. But even more interestingly, proportionality can hold between drastically different measurements, such as distance and time, when the rate of change is constant. And here we come down to yet one more translatable word from English to math:

- The phrase "per" denotes division and is usually referring to some rate of change.

Traveling at a constant speed of 32 miles (mi) per hour (hr) translates to

$$32 \, \frac{\text{mi}}{\text{hr}},$$

and we can think of this as a conversion factor

$$\frac{32 \text{ mi}}{1 \text{ hr}},$$

from hours to miles (provided the speed, which is a rate of change of distance over time, remains constant). For example, traveling for 2 hours at a speed of $32 \, \frac{\text{mi}}{\text{hr}}$ converts to

$$2 \, \text{hr} \times \frac{32 \text{ mi}}{1 \text{ hr}} = 64 \text{ mi}.$$

Inverting the conversion factor, we see that 6.4 miles, at a speed of 32 miles per hour, converts to

$$6.4 \, \text{mi} \times \frac{1 \text{ hr}}{32 \text{ mi}} = 0.2 \text{ hr},$$

which we can convert to minutes, if we wish, by another appropriate conversion factor:

$$0.2 \, \text{hr} \times \frac{60 \, \text{min}}{1 \, \text{hr}} = 12 \, \text{min.}$$

And these rates of change can be converted into other rates of change via conversion factors. For instance, we can, if we wish, convert 32 miles per hour into a number of feet per second, using conversion factors from miles to feet and from hours to seconds (recall that there are 5280 feet in every mile, 60 minutes in every hour and 60 seconds in every minute):

$$
\begin{aligned}
32 \, \frac{\text{mi}}{\text{hr}} &= 32 \, \frac{\text{mi}}{\text{hr}} \times \frac{5280 \, \text{ft}}{1 \, \text{mi}} \times \frac{1 \, \text{hr}}{60 \, \text{min}} \times \frac{1 \, \text{min}}{60 \, \text{sec}} \\
&= \frac{32 \times 5280}{60 \times 60} \, \frac{\text{ft}}{\text{sec}} \\
&= 46.93 \, \frac{\text{ft}}{\text{sec}}
\end{aligned}
$$

You can take conversion factors from the street into the gym with you. During my workouts I take my pulse. Rather than count the number of heartbeats over a whole minute, the usual process is to count the number of heartbeats in 10 seconds and multiply by 6 to get the number of beats per minute. Suppose that during a 10-second interval you count 11 beats, giving you a rate of $\frac{11 \, \text{beats}}{10 \, \text{sec}}$, which we can convert into the desired beats per minute:

$$\frac{11 \, \text{beats}}{10 \, \text{sec}} \times \frac{60 \, \text{sec}}{1 \, \text{min}} = 66 \, \frac{\text{beats}}{\text{min}}.$$

You might think it would be easier to count the number of beats in six seconds, and multiply by 10 to get the beats per minute, but accuracy would be at risk. No matter how careful you are, at each of the beginning or the end you might be off by a beat in your count. This translates into being off by possibly 6 x 2 = 12 beats per minute in your calculation if you did your count over 10 seconds, and 10 x 2 = 20 beats per minute if you did your count over six seconds instead, and an error like the latter is a much bigger deal.

You probably listen to music as you work out, and experiments have shown that people work out best to music when the tempo of the songs is the same as their pace on the treadmill. I favor listening to early Beatles' songs—why do these inspire me as I jog? I do a fast walk at a constant rate of 4.0 miles per hour. Let's use conversion factors to convert this to steps per minute. To do so, I need to know how long a step (or stride) of mine is. Putting a tape measure on the floor, I find it to be about two and half feet. Thus I walk at a pace of

$$4.0 \, \frac{\text{mi}}{\text{hr}} \; = \; 4.0 \, \frac{\text{mi}}{\text{hr}} \times \frac{5280 \, \text{ft}}{1 \, \text{mi}} \times \frac{1 \, \text{hr}}{60 \, \text{min}} \times \frac{1 \, \text{steps}}{2.5 \, \text{ft}}$$
$$= \; 141 \, \frac{\text{steps}}{\text{min}}$$

The tempo of many early Beatles songs (such as "I Want to Hold Your Hand") is right around 140 beats per minute, so I am right in step. Does your workout music keep you in step?

Conversion factors can be chained along as needed, but once again, keep the units present for proper canceling to ensure your answer is in appropriate units. It's not only good practice, but critical. Let me tell you two stories of how leaving out the units can take someone to the brink of disaster, and sometimes over.

A Deadly Lack of Math

A 43-year-old mother, Denise Melanson, passed away in 2006. She had been diagnosed six months prior with nasopharyngeal cancer, but that is <u>not</u> what she died from—her chemotherapy medication, Fluorouracil, was delivered intravenously over 4 hours as opposed to the prescribed 4 days. Although the mistake was discovered one hour after the end of the infusion, the overdose concentration resulted in multi-system failure for which there was no cure. Sadly, she died within a month of receiving the lethal dose.

Ms. Melanson's cancer was advanced, but treatable, and it was expected that, given appropriate treatment, she would recover. There were a number of similar reported fatalities between 2000 and 2007, and that does not count the undocumented or unreported serious medication errors added to this mix. Two reports were commissioned based on the horrific error. One of the reports stated:

You may have heard that listening to music can alleviate the intensity of pain. A study by Scandinavian researchers found that music truly makes pain easier to handle. They asked subjects (to whom they applied various degrees of discomfort) to compare listening to music to alleviate pain to other strategies as well, such as mental arithmetic. You might think that requiring people in pain to do simple mathematics would compound the issue, but NO—those who tried mental arithmetic actually felt less pain!

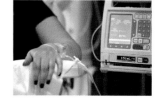

"For the following steps the nurse oriented the bag to read the multi-line label and used the calculator to check total dose, daily dose, and hourly dose calculations for internal consistency. The nurse did not use a piece of paper to write it out; the algebra was done in her head and the math on the calculator."

What makes even a bigger impression is that a second nurse checked the first nurse's calculations, or at least she tried to:

"... the checking RN (RN # 2) performed the calculation mentally and on a scrap of paper. ..."

Although it was determined that a number of factors came into play, one finding hits a nerve: "Nursing staff were required to complete a complex calculation to convert total dose in mg over 4 days to a rate in mL per hour." But the calculation was not so complex, if you know conversion factors.

The dosage of drug ordered was 5250 milligrams (mg) to be administered intravenously over 4 days. The delivery rate of the solution needed to be entered into the medication pump in milliliters per hour (mL/hr). The vials contained 500 mg of the drug in 10 mL of liquid. Therefore, the infusion rate in mL/hr needed would be

$$\frac{5250 \text{ mg}}{4 \text{ days}} = \frac{5250 \text{ mg}}{4 \text{ days}} \times \frac{10 \text{ mL}}{500 \text{ mg}} \times \frac{1 \text{ days}}{24 \text{ hr}} = 1.1 \frac{\text{mL}}{\text{hr}}.$$

What the nurses actually programmed the medication pump to infuse was 26.25 mL per hour of medication, exactly 24 times the required delivery rate—they left out the conversion factor from hours to days. Had they only written the calculation as we have, with the units, they would have arrived at the correct rate.

While the findings of a root cause analysis of this incident includes this faulty bedside calculation as only one of 15 errors that, together, caused this woman's death, could Ms. Melanson really have died from a lack of basic conversion skills among health care providers?

Exercises

Exercise 1.2.1. Suppose that x is related to y if $x > y$. Is the relation reflexive? symmetric? transitive? Explain your answers.

Exercise 1.2.2. Suppose that x is related to y if $x = y$. Is the relation reflexive? symmetric? transitive? Explain your answers.

Exercise 1.2.3. Suppose that person x is related to y if x likes y. Is the relation reflexive? symmetric? transitive? Explain your answers.

Exercise 1.2.4. Suppose that person x is related to y if x and y have a common friend. Is the relation reflexive? symmetric? transitive? Explain your answers.

Exercise 1.2.5. Suppose that y is proportional to x. If $y = 10$ when $x = 2$, what is the constant of proportionality?

Exercise 1.2.6. Suppose you drive at a constant speed, but then double your speed. How does the distance you cover in one hour compare, before and after your speed change?

Exercise 1.2.7. Do you think that the time you put into studying for an exam is proportional to the score you receive on the exam? Explain your answer.

Exercise 1.2.8. Do you think that the number of friends you can invite to a party is proportional to the amount of money you have to spend on the party? Explain your answer.

Exercise 1.2.9. Suppose you drive at a constant speed of 32 miles per hour. Explain why the distance you travel (in miles) is proportional to the time you travel (in hours).

Exercise 1.2.10. Suppose you drive at a constant speed of 32 miles per hour. Explain why the distance you travel (in feet) is proportional to the time you travel (in minutes).

Exercise 1.2.11. The time that passes in hours is proportional to the time that passes in seconds. What is the constant of proportionality?

Exercise 1.2.12. The time that passes in seconds is proportional to the time that passes in hours. What is the constant of proportionality?

Exercise 1.2.13. Do you think a person's height (in feet) is proportional to his or her weight (in pounds)? Explain your answer.

Exercise 1.2.14. For each of the following equations, decide whether x is proportional to y:
(a) $y = 2.2x$ (b) $y = 10.3x - 4.2$ (c) $1.1y - 6.2x = 0$ (d) $0.1y - 2.7x - 3.2 = 0$

Exercise 1.2.15. Convert each of the following temperatures to Fahrenheit:
(a) $15°C$ (b) $25°C$ (c) $0°C$ (d) $-10°C$ (e) $-40°C$

Exercise 1.2.16. Convert each of the following temperatures to Celsius:
(a) $32°F$ (b) $64°F$ (c) $0°F$ (d) $-10°F$ (e) $-40°F$

Exercise 1.2.17. An increase in $1°C$ corresponds to what temperature increase in Fahrenheit?

Exercise 1.2.18. An increase in $1°F$ corresponds to what temperature increase in Celsius?

Exercise 1.2.19. What is wrong with the following 2005 statement on global warming in the Antarctica, taken from `www.freerepublic.com/focus/news/1391368/posts`)?

"This region has shown dramatic and localized warming—around 2 degrees Celsius (35.6 degrees Fahrenheit) in the last 50 years," lead author of the study, Alison Cook, said in a statement."

Exercise 1.2.20. Every **byte** is made up of 8 bits (which are single digits, each a 0 or a 1). Are the number of bytes in a string of data proportional to the number of bits? Explain your answer.

Exercise 1.2.21. A **kilobit** (kbit) is equal to 1024 bits (see the previous exercise) and a **megabit** (Mbit) is equal to 1024 kilobits. A **megabyte** is equal to 1024 bytes. Write down the conversion factor from bits to megabits, and from bits to megabytes.

Exercise 1.2.22. Download speeds are usually given in kbps, kilobits per second, or in Mbps, megabits per second. If the download speeds are constant in each of the following scenarios, determine how fast a 320 megabyte file will download:

(a) 56 kbps (dial-up speed)   (b) 500 kbps (3G speed)   (c) 225 Mbps (cable modem speed)

Exercise 1.2.23. How far is 431 kilometers in miles?

Exercise 1.2.24. How far is 431 miles in kilometers?

Exercise 1.2.25. If 1 Euro is equal to 1.38228 U.S.D, what is 1 U.S.D worth in terms of Euros?

Exercise 1.2.26. If 1 Euro is equal to 1.38228 U.S.D, and 1 U.S.D is worth 1.12 CAN, what is 1 Euro worth in terms of Canadian dollars?

Exercise 1.2.27. Suppose you plan to charge $12.50 for a fully loaded 10-inch pizza. What should you charge for your world famous, giant 30-inch pizza?

Exercise 1.2.28. Suppose you plan to charge $14.99 for a fully loaded 10-inch pizza. What is the price per square inch of the pizza? (Round to the nearest penny.)

Exercise 1.2.29. Suppose that all pizzas that you make are squares.

(a) If your characteristic measurement is the diagonal d of the pizza, write down the area A of the pizza in terms of d (measure d in inches and A in square inches).

(b) If you plan on a price of 6 cents per square inch, write down the formula for determining the price of the pizza, in terms of the diagonal d.

Exercise 1.2.30. Are all trapezoids geometrically similar? Explain.

Exercise 1.2.31. Are all semicircles geometrically similar? Explain.

Exercise 1.2.32. Explain why all cubes are geometrically similar. What would you choose as a characteristic dimension?

Exercise 1.2.33. Are the following figures geometrically similar?

Exercise 1.2.34. Are the following figures geometrically similar?

Exercise 1.2.35. Suppose you have two sugar cubes, one with side length 1 cm and another with side length 2 cm. How would the weight of the second compare to the first? Why?

Exercise 1.2.36. Are all fish of a certain type (say sharks) geometrically similar or close to it? Explain your answer.

Exercise 1.2.37. Calculate your heart rate in beats per minute by

(a) counting the number of beats in 6 seconds,

(b) counting the number of beats in 10 seconds,

(c) counting the number of beats in 30 seconds, and

(d) counting the number of beats in 60 seconds.

How do your answers compare?

Exercise 1.2.38. Suppose your stride is 3 feet, and you plan to run at 5.5 miles per hour. What tempo of music should you listen to while running to get the maximum benefit?

Exercise 1.2.39. Measure your stride with a tape measure. Decide how fast you plan to run or jog on a treadmill. After calculating the ideal tempo of accompanying songs to listen to while running or jogging, pick a few appropriate songs.

Exercise 1.2.40. Suppose that you have a 16-year-old male relative who is itching like crazy from bug bites. You decide to give him an over-the-counter antihistamine. The problem is that he is very large for his age, and is off the scale given on the box, which looks like:

weight	dosage (every 6 hours)
under 24 lbs	2.5 ml
24–48 lbs	5.0 ml
48–95 lbs	10 ml

If the boy weighs 175 pounds, how much medication should you give him? How many tablespoons? You can use the fact that 1 tablespoon = 15 ml.

Exercise 1.2.41. In the formula for the fractional calorie goal formula (see Exercise 1.3.6), explain why the *contribution* to the units of the protein is proportional to the amount of protein.

Exercise 1.2.42. The *resolution* of a scanned image is the number of dots (or *pixels*) per inch that the image is divided into; it usually depends on the ultimate use of the image. For printed images, the resolution is often 300 dpi (*dots per inch*) or higher, for faxes it is 200 dpi, and for web display it is 72 dpi. Suppose that each dot requires 24 bits to represent its color.

(a) If the resolution is 72 dpi and the image is 4 inches by 5 inches, how many dots are there in the width and height of the image? What is the file size of the image?

(b) Explain why, for a fixed image, the file size is proportional to the *square* of the resolution.

(c) Explain why, for a fixed resolution, the file size is proportional to the area of the image.

(d) Suppose you have two versions of an image, one at a resolution of 72 dpi, suitable for the internet, and a higher resolution version at 360 dpi, for print. How do the file sizes compare?

Exercise 1.2.43. In a *grayscale* image, each dot requires only 8 bits to represent its grayscale. Suppose that each dot requires 24 bits to represent its color. If the resolution is 300 dpi and the image is 3 inches by 4 inches, what is the file size of the image?

Exercise 1.2.44. A $20 bill is approximately 0.0044 inches thick. By forming a conversion factor between a bill and inches, find the width, in feet, for a stack of 500 000 twenty dollar bills (which is one million dollars).

Exercise 1.2.45. The U.S. debt is approximately $17 528 000 000 000, that is, about 17.528 *trillion* dollars. Using the previous exercise, figure out how high a stack of twenty dollar bills the debt would be in miles. If you tried to wrap this stack of bills around the world (which has a circumference of 24 902 miles, how many times could you do so?

Exercise 1.2.46. Newton's Second Law of Motion states that for an object with mass m, if a net force of F acts on the object, the object undergoes an acceleration (that is, a change of velocity) of a with

$$F = ma.$$

(a) If the object's mass never changes, what does this say about the relationship between the net force and the acceleration?

(b) If the object's mass never changes, what happens to the object's acceleration if the net force is doubled?

(c) If the object's mass never changes, what can you say about the net force if the object's acceleration is cut in half?

(d) The acceleration due to gravity is constant for all objects on earth. The force of gravity of an object is its weight. If you have two objects, one having 10 times the mass of the other, what can you say about their relative weights?

Exercise 1.2.47. In the story of Denise Melanson, how many milliliters of medication did she receive over the 4 hours? How many milligrams of the drug Fluorouracil did she receive?

Exercise 1.2.48. In 1998, the Mars Climate Orbiter was sent on a mission to Mars. Unknown to NASA engineers, the orbiter was programmed to accept navigational information in the metric system, but the commands and data were sent from NASA in the imperial system. Instead of force being sent in "newtons," the numbers were sent in pounds. Look up details about the story online.

(a) What is the conversion factor between pounds and newtons?

(b) How far up from the planet was the orbiter supposed to orbit? How far up did the error place it?

(c) What was the end result for the orbiter?

(d) How costly was the conversion mistake?

And Now the Rest of the Story ...

Sometimes a lack of skill with conversion factors can lead to even more widespread peril. The chapter opening story of Air Canada flight 143 left off with the aircraft gliding, out of fuel, 41 000 feet up in the sky, over central Canada. How did this happen? A couple of years earlier, there was a legislated move by the Canadian government from the imperial system *(feet, miles, pounds, gallons, and so on) to the* metric system *(meters, kilometers, kilograms, liters, . . .), and many Canadians were inexperienced in the new metric system. As mentioned, the fuel gauges on the airplane were broken, and the maintenance crew had to do manual "drip" tests, much like the dipsticks used in cars to measure fluid levels. The pilot and copilot calculated that there was 7682 liters (L) of fuel in the plane during their stopover in Ottawa, and they knew that for the flight, they needed 22 300 kg of fuel.*

However, kilograms are a measure of weight, *and liters are a measure of* volume. *To convert from volume to weight, the pilots needed to use the* specific gravity *of the fuel, which is a conversion factor from volume to weight. It was known to be 1.77 pounds per liter (lbs/L), so the crew did the calculation:*

$$7682 \text{ L} \times \frac{1.77 \text{ lbs}}{1 \text{ L}} = 13\,567 \text{ kg}.$$

They then determined that 22 300 kg − 13 567 kg = 8703 kg of fuel was needed, so they ordered

$$8703 \text{ kg} \times \frac{1 \text{ L}}{1.77 \text{ lbs}} = 4916 \text{ L}$$

of fuel.

But notice that in the first calculation, the units were not carefully canceled; if the pilot had done so, they would have found that they had

$$7682 \text{ \cancel{L}} \times \frac{1.77 \text{ lbs}}{1 \text{ \cancel{L}}} = 13\,567 \text{ lbs}$$

of fuel, which is quite a bit different—the units are <u>not</u> the expected kilograms. Let's redo the calculation, converting from pounds to kilograms (there are 2.2046 pounds in a kilogram):

$$7682 \text{ \cancel{L}} \times \frac{1.77 \text{ \cancel{lbs}}}{1 \text{ \cancel{L}}} \times \frac{1 \text{ kg}}{2.2046 \text{ \cancel{lbs}}} = 6168 \text{ kg}.$$

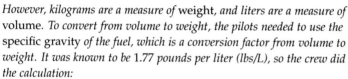

The proper conversion factor to use for specific gravity, in kilograms per liter, is

$$\frac{1.77 \cancel{lbs}}{1 \, L} \times \frac{1 \, kg}{2.2046 \cancel{lbs}} = 0.8029 \, \frac{kg}{L}.$$

They then would have determined that $22\,300 \, kg - 6168 \, kg = 16\,132 \, kg$ of fuel were needed, which requires

$$16\,132 \cancel{kg} \times \frac{1 \, L}{0.8029 \cancel{kg}} = 20\,092 \, L$$

of fuel—a HUGE difference from the 4,916 liters they ordered. That is why the plane ran dry in mid-air.

So what is the end of the story? Did lack of canceling of units in conversion factors do them in? Luckily, no. The skilled pilot and copilot managed to make an emergency landing at the Gimli airbase, which had been converted to a racetrack. As the plane hit the runway, two of the tires blew as the brakes were applied, and the nose of the plane scraped the tarmac, sparks flying. There were even some (very surprised!) people on the racetrack, but somehow the pilots managed to land the plane without hitting any pedestrians or racers. With only minor injuries (sustained in the frantic escape), everyone lived to tell the tale. Without exceptional piloting skills, the math lesson could have been much more costly!

3 | Review Exercises

Exercise 1.3.1. Suppose you go out of state and enter a restaurant where the sales tax is 6.5%. Your bill is $68.25 before tax.
(a) What is the sales tax on your bill?
(b) What tip should you leave? (Round to the nearest nickel.)
(c) What is the total of your bill, tax and tip included?

Exercise 1.3.2. Suppose you are about to purchase a new laptop with a sales price of $699.99. You notice a discount sticker of 15%. How much is the discount sticker worth? What is the discounted price of the laptop?

Exercise 1.3.3. If the sales tax is 6%, write down the formula for a function that calculates the sales tax on an item with price x dollars.

Exercise 1.3.4. If the sales tax is 8%, but all items in the store are discounted by 15% first, write down the formula for a function that calculates the total cost of an item with price x dollars.

Exercise 1.3.5. If the sales tax is 10%, and an item has a final cost of $54.00, what was the price of the item, before tax?

Exercise 1.3.6. The "Fractional Calorie Goal" function can be written as

$$\text{fraccgunits} = \frac{19 \times \text{carbs}}{175} + \frac{9 \times \text{fat}}{35} + \frac{16 \times \text{protein}}{175} - \frac{2 \times \text{fiber}}{25},$$

that is, it is the cgunits function, leaving off the rounding up.

(a) Calculate the fractional calorie goal units for the item in Exercise 1.1.33.

(b) Show that if you take two servings of an item, the number of fractional units is double that of one serving. Is the same true for regular units?

Exercise 1.3.7. Suppose that y is proportional to x. If $y = 10$ when $x = 15$, what is the constant of proportionality?

Exercise 1.3.8. The time that passes in seconds is proportional to the time that passes in hours. What is the constant of proportionality?

Exercise 1.3.9. If 1 Euro is equal to 1.38228 U.S.D, and 1 U.S.D is worth 1.12 CAN, what is 107 CAN worth in terms of Euros?

Exercise 1.3.10. Suppose you are making a recipe for another soup. You add in 4 cups of vegetable stock, 2 onions and 3 bags of carrots, and the recipe serves 8. Suppose you are having 7 people over for dinner and want to make just the right amount of soup for 7. How many cups of chicken stock, how many onions and how many bags of carrots should you use?

2 | *Making Sense of Your World with Statistics*

In 1998, and more extensively in 2001 and 2002, a big story broke in the British media about a link between MMR (Measles/Mumps/Rubella) vaccines and the development of autism, a disease that is potentially manifested by devastating behavioral and social problems for those afflicted. There had always been a fringe group of "anti-immunizers," who were against immunization of children, for a variety of reasons. Prior to 1998, lawyers had pressed ahead with cases, attempting to show that the MMR vaccine caused autism, but with little success. Dr. Andrew Wakefield, the lead researcher on the articles that drew the connection, championed their cause. There had long been rumblings of possible connections between the three-dose vaccine and autism, and here was scientific proof!

Indeed, Wakefield's research paper had all the hallmarks of a break-through in the area. It had been published in The Lancet, a top medical journal. An investigation of 12 children with autism-like problems showed that 8 of them developed symptoms soon after receiving the MMR vaccine. Data collected from a variety of sources all pointed to the same conclusion, even leading to a link with bowel disease, as well. Wakefield, basking in yet another TV interview, proclaimed that he recommended, based on his findings, that no one should be given the MMR vaccine, adding that he thought that separating the vaccines into three separate shots would be the safe way to proceed.

The media was all over the story—experimental science and statistics would save many a child from a debilitating disorder. Even Prime Minister Tony Blair was drawn into the fray—in light of the new findings, did he immunize his newborn son Leo with the MMR vaccine? The politician refused to confirm or deny. Public confidence in the use of the MMR vaccines fell drastically. In the UK alone, it dropped from 92% in 1998 to 84% in 2002, and even down to almost 60% in some areas in 2003. To be sure, actual cases of measles and mumps were on *the rise, but that seemed like a small price to pay for the avoidance of MMR-induced autism. Or was it?*

WE ARE CONSTANTLY INUNDATED WITH NUMBERS. Data, data everywhere. Your height, your weight, your love life, your bank account—are you normal? How do you judge? It seems to be impossible to make informed decisions without crunching, at least superficially, the numbers that surround you. You expect that you can't control everything, that there will be some variation in whatever you measure. But certainly there are trends and there are patterns. If only you could separate out the truth from the noise, things would be clearer, and life would be simpler. Making sense out of data is exactly why **statistics** exists. While the field is large and deep, we can wade into the shallow end to explore some of the relevant notions and concepts, and even learn a few useful calculations along the way.

2.1   *Summarizing Data with a Few Good Numbers*

The news is filled with streams of numbers. For example, here is a listing of the exchange rates for U.S. dollars over the last 12 months:

$$1.1074, 1.1138, 1.0636, 1.0620, 1.0427, 1.0303,$$
$$1.0530, 1.0272, 1.0518, 1.0368, 1.0075, 1.0160.$$

The numbers of course show some variation, up and down, but certainly they are somewhat close together. We humans are overwhelmed by large streams of numbers (computers aren't!) and like to condense it down to a few representative figures that summarize such a list. Here is a guiding principle that will show up often in this book:

Life Lesson: When confronted with a lot of data, people like to condense the numbers into a smaller descriptive list, even at the expense of losing some information. **Compression of a large quantity of information into a smaller amount is always desirable and appreciated.**

Numbers that Describe

Descriptive statistics are numbers that summarize data, in some way. Let's start with a single number. Here, there are a few common choices.

Definition 2.1.1. Given a list of numbers,

- the **mean** is the **average** of the numbers,

- the **median** is the middle value, when you order the list, if there is an odd collection of numbers, and the average of the middle two if there is an even collection of numbers, and

- the **mode** is the most common number in the list (a list could have more than one mode, or no mode if no number repeats).

Note that all we can say definitively is that *approximately* half the items are less than or equal to the median and half greater than or equal to it, depending on the number of items in the data list and the number of repetitions of the median value. For example, the median of $1, 2, 5, 6, 9$ is 5 but, as the list is odd, more than half the numbers are less than or equal to 5, and fewer than half are greater than or equal to 5—but it is pretty close. Similarly for the data list $1, 2, 5, 5, 6, 9$, with median 5.

Let's consider a monthly list of currency exchange rates for U.S. dollars in terms of Canadian dollars:

$$1.1074, 1.1138, 1.0636, 1.0620, 1.0427, 1.0303,$$
$$1.0530, 1.0272, 1.0518, 1.0368, 1.0075, 1.0160.$$

The sum of all 12 values turns out to be 12.6121, and by dividing by 12, we find that the mean is 1.0510. When we reorder the numbers, we get the list:

$$1.0075, 1.0160, 1.0272, 1.0303, 1.0368, 1.0427,$$
$$1.0518, 1.0530, 1.0620, 1.0636, 1.1074, 1.1138$$

There is no middle number, as there are 12 items; we find the median by averaging the middle two numbers—the median is

$\frac{1.0427+1.0518}{2} = 1.0472$. Finally, no number repeats, so there is no mode. The mean and median are close together, and both seem like good central measures of the data.

Let's look at another example. Suppose we start with a list of grades on test for a small math class of 9 students:

$$67, 41, 83, 68, 55, 59, 68, 95, 62.$$

The mean is the average—that is, we add up all of the numbers and divide by the total number of items (which happens to be 9 in this case):

$$\text{mean} = \frac{67 + 41 + 83 + 68 + 55 + 59 + 68 + 95 + 62}{9} = \frac{598}{9} \approx 66.4.$$

The median is the middle number, and to find this, we need to order the list, from smallest to highest or vice versa (it doesn't matter which):

$$41, 55, 59, 62, 67, 68, 68, 83, 95.$$

The list has 9 items; the fifth number, 67, is right in the middle, so

$$\text{median} = 67.$$

Finally, the number that appears most often is 68 (it appears twice, and every other number appears a fewer number of times, so

$$\text{mode} = 68.$$

All of these seem a reasonable one-number description of the list of 9 numbers, each some sort of "central" number of the list.

Sometimes data is presented in a **frequency table**, which lists the *different* data values together with the number of times each appears in the data. This is particularly useful when various values appear more than once (in such a case, the frequency table is again a compressed view of the data, and this compression is again appealing to people). For example, if the heights in inches for members of a basketball team are

$$76, 78, 76, 82, 81, 82, 76, 76, 85, 76, 78, 82, 85$$

then we can form a frequency table that is easier to carry around (see Table 2.1).

height (in.)	76	78	81	82	85
frequency	5	2	1	3	2

Table 2.1: Example of a frequency table.

You can do the same calculations from the frequency table that you can do from the original data list, recognizing that the frequency table is only a condensed form of the data, with no information lost (you could, of course, reconstruct the original data list from the frequency list, if you chose to do so). So, for example, we can calculate the mean of the basketball heights by averaging all of the 13 numbers; alternatively, from the frequency table, we see that the sum of all the heights is

$$5 \times 76 + 2 \times 78 + 1 \times 81 + 3 \times 82 + 1 \times 85 = 948,$$

so the mean is $948/13 = 72.9$. The median would be the seventh number in the ordered list, which we can see from the frequency table is 78. Finally the mode is trivial to read off from the frequency table, being 76.

You'll find that in many places where lots of data is given, the presenter summarizes by giving one of the mean, median or mode. Which is more meaningful? That is a difficult question to answer, but you should always be aware that one or the other may not be representative of the data. For example, suppose we were looking at the income of residents in an area of town that is a mix of many low income earners and a few high income earners. For argument's sake, suppose that there are 90 residents who earn only $20 000 annually, but 10 who earn $1 000 000. The frequency table is:

income (dollars)	20 000	1 000 000
frequency	90	10

The mean income is

$$\frac{90 \times \$20\,000 + 10 \times \$1\,000\,000}{100} = \$118\,000,$$

which would be a misleading figure for the "central" income, as no one earns near this amount—a few earn more, but most earn much, much less! The median, $20 000, would be a better choice of a central number for the data, as most people in the region earn this amount.

The median, which is the value for which about half of the data is below and half of the data is above, is a useful measure, as it gives you an idea of where the "center" is, so if you have a score above the median, you would know that at least half of the other scores would sit below yours—the median tells you something about your position in the data (but <u>not</u> the value of your score). But to get a better idea of your ranking with respect to others, the idea of a median can be extended.

Mathematics spends a lot of time extending notions to more general ones.

Definition 2.1.2. For a number p between 0 and 100, the pth **percentile** is the value x for which (approximately) $p\%$ of the data are less than or equal to x.

The median is simply the 50th percentile.

Note the use of the word *approximately* in the definition of percentile, as clearly, just based on the number of values in the data, you may not be able to find an exact value (just as the median is only *approximately* the value for which half the numbers are below). There is no universally accepted precise definition of percentile—it varies slightly from textbook to textbook.

But based on our understanding from Chapter 1 of percentages, let's try an example. Suppose we have the following list of 109 test scores, all out of 50:

21	19	30	15	8	35	37	25	25	40	39	35	44	41	21
43	38	43	22	36	41	37	40	18	16	44	26	18	23	25
29	37	44	29	17	42	25	25	36	36	43	43	45	14	37
21	15	31	27	27	30	42	34	29	43	34	35	32	41	30
38	39	33	28	35	33	38	30	18	31	34	50	28	30	29
30	40	30	28	34	41	27	45	32	34	37	38	35	25	36
36	32	27	34	32	42	31	33	37	27	26	39	35	32	40
46	28	42	29											

First, suppose your score was 41. What is your percentile? The best way to begin is to reorder the list:

8	14	15	15	16	17	18	18	18	19	21	21	21	22	23
25	25	25	25	25	25	26	26	27	27	27	27	27	28	28
28	28	29	29	29	29	29	30	30	30	30	30	30	30	31
31	31	32	32	32	32	32	33	33	33	34	34	34	34	34
34	35	35	35	35	35	35	36	36	36	36	36	37	37	37
37	37	37	38	38	38	38	39	39	39	40	40	40	40	41
41	41	41	42	42	42	42	43	43	43	43	43	44	44	44
45	45	46	50											

As there are 109 numbers in the data set, the median is the middle number, which is the 55th, that is, 33.

From the ordered list, we see that there are 93 scores less than or equal to 41, that is,

$$\frac{93}{109} = 0.853 = 85.3\%$$

so you are at the 85th percentile.

Suppose, on the other hand, we want to find the 68th percentile, that is, the number such that 68% of the data is less than or equal to it. Now 68% of 109 is

$$0.68 \times 109 = 74.12$$

so we are looking for the 74.12th number in the ordered list of 109 numbers, so we clearly need to take either the 74th or 75th number.

- If we take the 74th number, 36, we get $74/109 = 0.679 = 67.9\%$ of the numbers less than or equal to it.

- If we take the 75th number, 37, we get $75/109 = 0.679 = 68.8\%$ of the numbers less than or equal to it.

So we have a choice to make. We can either

- take 36 as the 68th percentile, as it gives the closest percentage to the desired 68%, though it is a little short, or

- move up to 37 as the 68th percentile, as it is the first value that has *at least* 86% of the numbers less than or equal to it.

There is yet another choice. Since the difference in percentages less than or equal to each of 36 and 37 is $68.8 - 67.9 = 0.9$ and 68 is $68.0 - 67.9 = 0.1$ up from the lower percentage, 68 is $\frac{0.1}{0.9} = 0.11.1 = 11.1\%$ along the way up from 67.9 to 68.8. So it would make sense to say the 68th percentile is 11.1% along the way from 36 to 37. As the distance between 36 and 37 is 1, we take as our 68th percentile

$$36 + 0.111 \times 1 = 36.111.$$

While this might seem like an odd choice, it is really the way we define the median, with the averaging of the two central numbers, if the number of data is even.

(This method of estimating a certain value between points in a list is called **linear interpolation**. The idea is that if you want to move $x\%$ from a to b, where $a \leq b$, then you take the value $a + (x/100) \times (b - a)$. The value $x/100$ is x as a decimal, and $b - a$ is the length of the entire distance from a to b. Linear interpolation is a very common method for estimating values between data points—you calculate what percentage you move away from the left end's x-value, and then move vertically the same *percentage* from the left end's y-value toward the right end.)

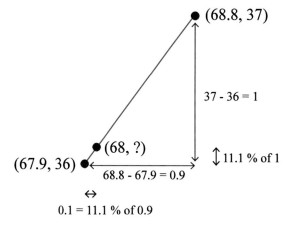

Figure 2.1: Linear interpolation between two points.

Percentiles are the way SAT test scores are evaluated, as it gives you the clearest indication of how you fared compared to others who took the examination at the same time (actual scores can vary somewhat from year to year, depending on the difficulty of the year's examination).

We have discussed single numbers that describe data, but of course a single number can't encapsulate the full scope of the data. For example, let's compare two data lists:

$$10, 10, 10, 10, 10, 10, 10, 10, 10, 10$$

and

$$5, 5, 5, 10, 10, 10, 10, 15, 15, 15.$$

A quick calculation will show that both lists of data have the same mean, median and mode, namely 10 in both cases. But clearly the data is drastically different. The first has all numbers at the mean/median, while the second has most values away from the mean/median/mode but the numbers are equally balanced on either side.

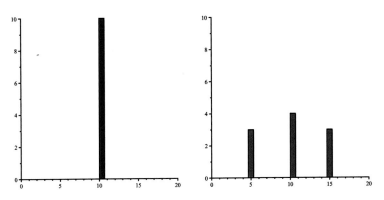

Figure 2.2: Two data sets with the same mean and median—the height denotes the number of times the value occurs in the data set. In the next chapter we'll discuss more about the visual representation of the data.

Giving a High Five

Clearly a single measure like the mean or median cannot distinguish between these. So we need to find a way to distinguish them, perhaps with other sets of numbers.

One approach is based on the median and percentiles. It uses a set of five numbers based on the data:

- the minimum value,

- the **first quartile** or **lower quartile** Q_1,

- the median Q_2,

- the **third quartile** or **upper quartile** Q_3, and

- the maximum value.

(A **quartile**, arising from the word *quarter*, is 25% of the data, so the first quartile Q_1 is the 25th percentile, the second quartile Q_2 is the $25 + 25 = 50$th percentile, that is, the median, and the third quartile Q_3 is the $25 + 25 + 25 = 75$th percentile.) This list of numbers is called, not surprisingly, the **five-number summary** of the data list. As there were some choices for the definition of the percentiles, we take as our first and third quartiles the medians of the data lists below and above the median of the whole data list (the definition of the median of a list was not up for discussion!).

The first and last numbers of the five-number summary give the range for the data, and the middle three numbers give the values where approximately 25%, 50% and 75% of the values lie to the left.

So for $10, 10, 10, 10, 10, 10, 10, 10, 10, 10$, we easily see that the minimum, the first quartile Q_1, the median (i.e., the second quartile Q_2), the third quartile Q_3 and the maximum are all 10, so the five-number summary is

$$10, 10, 10, 10, 10.$$

For the second data list, $5, 5, 5, 10, 10, 10, 10, 15, 15, 15$, we find that

- the minimum is 5,

- the median is 10, occurring between the middle two 10's

- the first quartile Q_1 is the median of the bottom half of the list, $5, 5, 5, 10, 10, 10$, which is $(5 + 10)/2 = 7.5$,

- the third quartile Q_1 is the median of the bottom half of the list, $10, 10, 10, 15, 15, 15$, which is $(10 + 15)/2 = 12.5$, and

- the maximum is 15.

Thus the five-number summary for the data $5, 5, 5, 10, 10, 10, 10, 15, 15, 15$ is

$$5, 7.5, 10, 12.5, 15,$$

which is quite different from

$$10, 10, 10, 10, 10,$$

the five-number summary for the data $10, 10, 10, 10, 10, 10, 10, 10, 10, 10$.
Let's consider a less trivial example, the 109 scores.

```
21  19  30  15   8  35  37  25  25  40  39  35  44  41  21
43  38  43  22  36  41  37  40  18  16  44  26  18  23  25
29  37  44  29  17  42  25  25  36  36  43  43  45  14  37
21  15  31  27  27  30  42  34  29  43  34  35  32  41  30
38  39  33  28  35  33  38  30  18  31  34  50  28  30  29
30  40  30  28  34  41  27  45  32  34  37  38  35  25  36
36  32  27  34  32  42  31  33  37  27  26  39  35  32  40
46  28  42  29
```

By ordering the values,

```
 8  14  15  15  16  17  18  18  18  19  21  21  21  22  23
25  25  25  25  25  25  26  26  27  27  27  27  27  28  28
28  28  29  29  29  29  29  30  30  30  30  30  30  30  31
31  31  32  32  32  32  32  33  33  33  34  34  34  34  34
34  35  35  35  35  35  35  36  36  36  36  36  37  37  37
37  37  37  38  38  38  38  39  39  39  40  40  40  40  41
41  41  41  42  42  42  42  43  43  43  43  43  44  44  44
45  45  46  50
```

The minimum and maximum values are 8 and 50, respectively.
As there is an odd number of values, 109, the median Q_2 is the
middle value, the 55th, namely, 33 (underlined and in bold).

```
 8  14  15  15  16  17  18  18  18  19  21  21  21  22  23
25  25  25  25  25  25  26  26  27  27  27  27  27  28  28
28  28  29  29  29  29  29  30  30  30  30  30  30  30  31
31  31  32  32  32  32  32  33  33  33  34  34  34  34  34
34  35  35  35  35  35  35  36  36  36  36  36  37  37  37
37  37  37  38  38  38  38  39  39  39  40  40  40  40  41
41  41  41  42  42  42  42  43  43  43  43  43  44  44  44
45  45  46  50
```

That breaks up the list into two parts, the upper part before the median,

8	14	15	15	16	17	18	18	18	19	21	21	21	22	23
25	25	25	25	25	25	26	26	27	27	27	27	27	28	28
28	28	29	29	29	29	29	30	30	30	30	30	30	30	31
31	31	32	32	32	32	32	33	33						

and the lower part

										34	34	34	34	34
34	35	35	35	35	35	35	36	36	36	36	36	37	37	37
37	37	37	38	38	38	38	39	39	39	40	40	40	40	41
41	41	41	42	42	42	42	43	43	43	43	43	44	44	44
45	45	46	50											

each of size 54. As 54 is even, in each list we get its median by averaging the 27th and 28th values, namely, $Q_1 = \frac{27+27}{2} = 27$ and $Q_2 = 33$ and $Q_3 = \frac{38+39}{2} = 38.5$. Thus the five-number summary of the data is

$$8, 27, 33, 38.5, 50.$$

The **frequency plot** of the original data is shown in Figure 2.3 and a **box plot** (or the "box and whisker" plot) of the five-number summary is shown in Figure 2.4 (the left edge and right edges are the minimum and maximum of the data, the left edge of the box is the first quartile, the right edge of the box is the third quartile, and finally the line in the middle of the box is the median). We'll talk more about graphical representation of numbers in the next chapter.

Sometimes box plots are drawn vertically rather than horizontally, but the process is the same.

The five-number summary provides an idea of the *spread* or *variability* of the data, much more than just the median. For example, the 109 test scores have a median of 33, which only tells you that (about) half the class scored below 33 and half above. So if you scored 41 on the test, you would have scored in the top half, but that would be all you could say from just knowing the median. But from the five-number summary, $8, 27, 33, 38.5, 50$, you would know that the third quartile is at 38.5, so that your

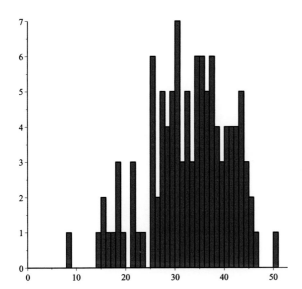

Figure 2.3: Frequency plot.

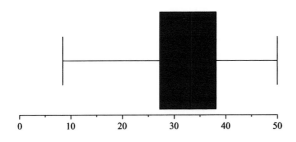

Figure 2.4: Box plot of the frequency distribution in Figure 2.3.

score of 41, being above it, would be in the top 25% of the class—much more information.

In fact, the **range** of a data set is defined to be the difference between the maximum and minimum values, and the **interquartile range** is the difference between the third and first quartiles. So, for example, in our 109 score data set with five-number summary $8, 27, 33, 38.5, 50$, the range is $50 - 8 = 42$ and the interquartile range is $38.5 - 27 = 11.5$. So all of the scores are between 8 and 50 (inclusive), an interval of size 42, and approximately the middle 50% of the scores are between 27 and 38.5, an interval of size 11.5—this is the "boxy" part of the box plot in Figure 2.4 (the vertical lines indicate the minimum and maximum values, while the "whiskers"—the horizontal lines—indicate the lower and upper 25% of the scores).

Outside the Norm

The interquartile range is useful for describing **outliers**, that is, data values that are very unusual—they are either way too small or way too large. For example, if in a hockey season you had all forwards except one with goals scored between 0 and 50, with one player scoring 75 goals in the season, that value would be so unusual as to be called an outlier. But how do we precisely define what is "unusual," what constitutes an outlier? After all, with a lot of data, we might expect some values to be small and some to be large. The usual definition is that an outlier is the following:

Definition 2.1.3. For a given data collection, let Q_1 and Q_3 be the lower and upper quartiles and let $IQR = Q_3 - Q_1$ be the interquartile range. Then a value less than $Q_1 - 1.5 \times IQR$ or greater than $Q_3 + 1.5 \times IQR$ is called an **outlier** for the data.

When a data value is an outlier, it sometimes indicates something out of the ordinary is going on with that value. A list of the Major League Baseball's home run leaders from 1962 to 2001 is shown in Table 2.2 (the frequency plot is shown in Figure 2.5).

All seasons consisted of 162 games, except for strike-shortened seasons in 1972 (156 games), 1981 (111 games), 1994 (117 games) and 1995 (144 games). I'll remove these years from consider-

Figure 2.5: Baseball player Barry Bonds.

ation. You can check that the remaining 36 home run numbers have a median $Q_2 = 46.5$, lower quartile $Q_1 = 43$, upper quartile $Q_3 = 49.5$, so that the five-number summary is $36, 43, 46.5, 49.5, 73$. The interquartile range is $49.5 - 43 = 6.5$, so that any value smaller than $43 - 1.5 \times 6.5 = 33.25$ or greater than $43 + 1.5 \times 6.5 = 59.25$ is an outlier. The number of home runs hit by Mark McGwire in 1998 (70 HR) and 1999 (65 HR) and those hit by Barry Bonds (see the top plot in Figure 2.6) in 2001 (73 HR) are clearly outliers (the box plot shown at the bottom of Figure 2.6 is of the data minus the outliers, with the outliers shown as x's outside the whiskers). It has come out that both of these players, despite denials, used performance-enhancing drugs (including steroids), so these statistical outliers were indeed outside the ordinary, or what should have been the ordinary!

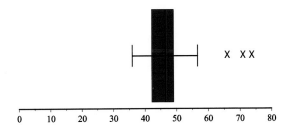

Figure 2.6: Histogram and box plot of leading home runs 1962–2001, with outliers as x's.

Life Lesson: It is sometimes hard to tell *outliers* from *outed liars*.

Spread 'Em

While the use of the five-number summary is one way to describe the spread of the data, there is another common one, related to the mean, called the standard deviation.

Definition 2.1.4. Suppose that $x_1, x_2, \ldots, x_n$ is a list of data values, with mean $\bar{x}$. Then the **variance** is defined to be the sum of the squares of the differences between each of the data values and the mean, $\bar{x}$, divided by n, the number of values. The **standard**

Year	Home Run Leader	Home Runs (HR)
1962	Willie Mays	49
1963	Harmon Killebrew	45
1964	Harmon Killebrew	49
1965	Willie Mays	52
1966	Frank Robinson	49
1967	Harmon Killebrew, Carl Yastrzemski	44
1968	Frank Howard	44
1969	Harmon Killebrew	49
1970	Johnny Bench	45
1971	Willie Stargell	48
1972	Johnny Bench	40
1973	Willie Stargell	44
1974	Mike Schmidt	36
1975	Mike Schmidt	38
1976	Mike Schmidt	38
1977	George Foster	52
1978	Jim Rice	46
1979	Dave Kingman	48
1980	Mike Schmidt	48
1981	Mike Schmidt	31
1982	Reggie Jackson, Gorman Thomas	39
1983	Mike Schmidt	40
1984	Tony Armas	43
1985	Darrell Evans	40
1986	Jesse Barfield	40
1987	Andre Dawson, Mark McGwire	49
1988	Jose Canseco	42
1989	Kevin Mitchell	47
1990	Cecil Fielder	51
1991	Jose Canseco, Cecil Fielder	44
1992	Juan Gonzalez	43
1993	Barry Bonds, Juan Gonzalez	46
1994	Matt Williams	43
1995	Albert Belle	50
1996	Mark McGwire	52
1997	Ken Griffey	56
1998	Mark McGwire	70
1999	Mark McGwire	65
2000	Sammy Sosa	50
2001	Barry Bonds	73

Table 2.2: Home run leaders in Major League Baseball (MLB) (1962–2001).

deviation is the square root of the variance.

To calculate the standard deviation, it is often helpful (at least when there aren't too many data values) to create a table with three columns, with the data down the left column (which we label x_i). The sum of that column, divided by the number of entries, is the mean, $\bar{x}$. The middle column, which we label $x_i - \bar{x}$, are the *deviations* (that is, differences) of each of the data values from the mean. The third column, labeled $(x_i - \bar{x})^2$, is the square of the deviations (that is, the squares of the numbers in the second column). The sum of that last column, divided by the number of data values, is the variance, and the square root of that is the standard deviation.

For example, for the data set

$$67, 41, 83, 68, 55, 59, 68, 95, 62$$

we found earlier that the mean $\bar{x}$ was 66.4. We form the following table:

data x_i	deviations $x_i - \bar{x}$	squared deviations $(x_i - \bar{x})^2$
67	0.6	0.36
41	-25.4	645.16
83	16.6	275.56
68	1.6	2.56
55	-11.4	129.96
59	-7.4	54.76
68	1.6	2.56
95	28.6	817.96
62	-4.4	19.36
598	0.4	1948.24

Table 2.3: Example of the calculation of squared deviations.

The sum of each column is shown at the bottom. The sum of the first column, 598, divided by the number of data values, 9, is 598/9, which is approximately 66.4—this is the mean (or average value). The next column takes the mean away from each data value. The third column is the square of each number in the middle column. The sum of this column, 1948.24, divided by the number of data values, 9, is the variance, which in this case is

$1948.24/9 = 216.47$. The square root of this, $\sqrt{216.47} = 14.7$, is the standard deviation of the data.

Why is the standard deviation a measure of the spread of the data? Well, the mean is a measure of the center of the data. When we take the differences of the data values each from the mean (these are the *deviations* from the mean), we get a measure of how far each is from the mean—if the value is negative, the number is to the left of the mean, and if it is positive it is to the right. All we want to measure is how *far* the number is from the mean, not whether it is to the right or the left, so to make the deviations positive, we square them—the further it is away from the mean, the larger the squares will be. Summing all of these squares will give us a positive number (or possibly 0, in the unlikely event that all the data are equal to the mean) that measures the *total* spread about the mean. Then dividing by the number of data values gives some average spread. But because we squared the deviations, it has the "wrong units," as described in our discussions in the previous chapter, so to fix this, we take the square root, and arrive at the standard deviation as our measure of spread.

If the original data was in units of say "points," then so will be the deviations, and the squares of the deviations will be in "points × points," or "points^2." Dividing by the number of data values (which has no units), we see that the variance also has units "points^2". But our measure of spread should have the same units as our original data, so to fix this, we take the square root which has units "$\sqrt{\text{points}^2}$," which is "points" again.

Note that within one standard deviation, 14.7, of the mean, 66.4, that is between $66.4 - 14.7 = 51.7$ and $66.4 - 14.7 = 81.1$, much of the data lies—only three values out of the 9, namely, 41, 83 and 95 lie outside this range. So 6 out of the 9, about 67% of the data, lie within one standard deviation of the mean. And within two standard deviations of the mean, that is, in the interval from $66.4 - 2 \times 14.7 = 37.0$ and $66.4 - 2 \times 14.7 = 95.8$ <u>all</u> of the data lie.

The example is rather small, so let's look back at the list of 109 test scores. We won't write out the table (as it will have many rows), but using a calculator (or a program like *Excel*) we find that the mean and standard deviation are 32.4 and 8.2, respectively. Within one standard deviation of the mean, that is, between $32.4 - 8.2 = 24.2$ and $32.4 + 8.2 = 40.6$, we see from the sorted data

8	14	15	15	16	17	18	18	18	19	21	21	21	22	23
25	25	25	25	25	25	26	26	27	27	27	27	27	28	28
28	28	29	29	29	29	29	30	30	30	30	30	30	30	31
31	31	32	32	32	32	32	33	33	33	34	34	34	34	34
34	35	35	35	35	35	35	36	36	36	36	36	37	37	37
37	37	37	38	38	38	38	39	39	39	40	40	40	40	41
41	41	41	42	42	42	42	43	43	43	43	43	44	44	44
45	45	46	50											

that 74 of the 109, or about 68%, fall in this interval. Within two standard deviations of the mean, that is, between $32.4 - 2 \times 8.2 = 16.0$ and $32.4 + 2 \times 8.2 = 48.8$, 104 of the numbers, or about 95%, lie in the interval—most of them. And within three standard deviations of the mean, that is, between $32.4 - 3 \times 8.2 = 7.8$ and $32.4 + 3 \times 8.2 = 57.0$, all 109 of the numbers (100%) are in the interval.

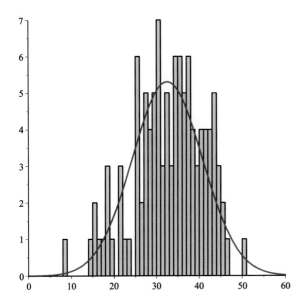

Figure 2.7: A bell-shaped curve.

It may be surprising that the percentage of numbers that lie within one, two or three standard deviations of the mean were so close in both examples—67%, 100%, 100% and 68%, 95%, 100%.

58

But that happens frequently with data sets, especially large ones that comes from measurements:

> The **68 − 95 − 99.7 rule**: For data that arise as measurements that follow a 'bell-shaped" pattern, you can expect about 68% of the data to lie within one standard deviation of the mean, about 95% within two standard deviations of the mean, and about 99.7% within three standard deviations of the mean.

It's worth mentioning that while 68 − 95 − 99.7 rule holds approximately for many distributions of numbers, it holds exactly for those that follow a *certain* bell-shaped curve called the **normal curve**. Its mathematical description can look a bit intimidating, but the shape should be familiar to you:

A normal distribution is *bell-shaped*, and often grades (such as SAT scores) are expected to follow such a curve.

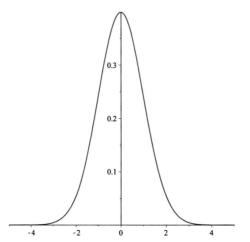

Figure 2.8: Plot of the standard normal curve.

Normal curves can have any mean and any standard deviation, but the one shown, the **standard normal curve**, has mean 0 and standard deviation 1. For those data that have a **normal distribution**, that is, their frequencies of different values follow a normal curve,

For a normal curve, the mean, median and mode are all equal, and the curve is symmetric about its center.

• within one standard deviation of the mean, 68% of the values lie,

- within two standard deviations of the mean, 95% of the values lie, and

- within three standard deviations of the mean, 99.7% of the values lie.

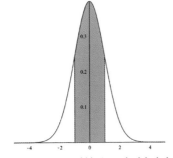

68% of values are within 1 standard deviation

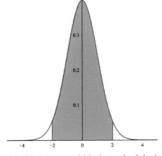

95% of values are within 2 standard deviation

Figure 2.9: One, two and three standard deviations away from the mean in the normal curve.

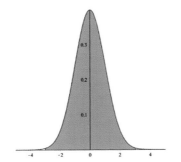

99.7% of values are within 3 standard deviation

Many distributions, especially those that arise from measurements and those that arise from sampling (see the next section!), turn out to be close to a normal distribution, and that is why the $68 - 95 - 99.7$ rule holds often, and why the standard deviation is such a useful measurement of the spread of data!

Getting the Right Angle on Sports

We end this section by showing that we can always concoct new descriptive statistics, and sometimes they can be very useful.

Sports is rife with statistics—baseball has home runs, runs batted in, batting average; hockey has goals, assists, goals against average; basketball has points per season, points per game, assists, rebounds, and so on. All of these are used to compare players or teams against one another. But is there a statistic to garner how much a player is worth to a team? In the fall of 2011, Sydney Crosby, the star hockey player for the Pittsburgh Penguins, was put out of action, after a pair of concussions, with the worst being Washington Capital's David Steckel's blindsided shoulder hit to Sidney's head. The loss of Crosby affected not only the Penguins, but hockey itself. Is there a way to quantitatively evaluate just how costly the loss of Sidney was to the Penguins?

Figure 2.10: Pittsburgh Penguin's Sydney Crosby (a.k.a. Syd the Kid).

The application of statistics and mathematics to baseball (and more generally, to sports) is called **sabermetrics**. The father of sabermetrics, statistician Bill James, proposed a formula for how to estimate the percentage of games won by a baseball team in a season from the number of runs scored (S) and the number of runs allowed (A):

$$\text{winning percentage} = \frac{S^2}{S^2 + A^2}.$$

You might recognize the denominator of the fraction as looking like the Pythagorean formula for calculating the length of the square hypotenuse from the sum of the squares of the other two sides. Because of this similarity, the formula is often called the **Pythagorean winning percentage**. If you examine the formula, you'll see that increasing the runs scored, or decreasing the runs allowed, increases the winning percentage, as it should.

On the surface it sounds like it would be impossible to predict the percentage of games won from the runs scored and runs allowed, as in any given game, you could score many more runs than you need to win, and yet get only one win, and in another game you could lose badly and still it's only one loss. Yet we are talking not about a single game, but the sum total of all the games in a long season, and this is where statistics comes to the fore. Perhaps surprisingly, the formula actually works quite well for baseball teams, giving results that are almost always within 5% of the true value.

The formula can be carried over to many other goal- scoring team sports, such as basketball, football and hockey. So here is a Pythagorean descriptive statistic for predicting the number of games won by a hockey team during an 82-game season, based on the number of goals scored (S) and goals allowed (A):

$$\text{games won} = 82 \times \frac{S^2}{S^2 + A^2}.$$

Tie games are ignored, but the formula seems to work well enough as is. For example, in the 2006–07 season, the Penguins scored 267 goals and allowed 240, with 47 games won. The Pythagorean formula would predict, just based on the goals scored and allowed, that they would have ended up winning

$$82 \times \frac{267^2}{267^2 + 240^2} = 82 \times \frac{7921}{14321} \approx 45$$

games, so we are pretty close with the prediction.

Now, how did Sydney's 120 points contribute to the Penguin's success? Well, without those, the team would have scored 120 fewer goals, and the Pythagorean formula suggests the team would only have won about

$$82 \times \frac{147^2}{147^2 + 240^2} \approx 22$$

The Pythagorean winning percentage is only an approximation of Sydney's contribution to the team—of course, if Sydney is not playing, other players would get more playing time, and undoubtedly at least some other goals would be scored.

games, so having Sidney on the ice added about 25 wins to the season! In the 2006–07, 2008–09, 2009–10 seasons (I didn't include Sidney's rookie year nor the ones when he played significantly less than 82 games), the formula predicts that in each of these seasons, just based on his offensive output, Sidney added at least 20 wins per season to the Penguins. The Pythagorean descriptive statistics shows that in hockey's valuable commodity of wins, Syd the Kid is a top broker—his scoring is the difference between the Penguins being at the ceiling or the basement of the league.

Exercises

Exercise 2.1.1. For the data $4, 6, 4, 3, 2, 5, 4$ find the mean, median and mode.

For Exercises 2.1.2 to 2.1.11, use the following data, taken from assignment scores in a math class:

$$66, 72, 55, 66, 80, 66, 57, 52, 57, 63, 55, 57, 75, 75, 70, 78, 57, 48, 59, 64, 75, 39, 70, 60, 72, 67, 60, 77.$$

Exercise 2.1.2. Write down the frequency table.

Exercise 2.1.3. Find the mean, median and mode.

Exercise 2.1.4. (a) To what percentile does the value 67 correspond? (b) The value 77 is at what percentile?

Exercise 2.1.5. What data value is at the 25th percentile?

Exercise 2.1.6. Use linear interpolation to find the 70th percentile.

Exercise 2.1.7. (a) Draw a frequency plot of the data. (b) Draw a box plot of the data.

Exercise 2.1.8. (a) Find the first and third quartiles. (b) Find the range and the interquartile range.

Exercise 2.1.9. Find the five-number summary.

Exercise 2.1.10. Are there any outliers? Explain.

Exercise 2.1.11. The standard deviation of the scores is 9.72.
(a) What proportion of the scores lie within one standard deviation of the mean?
(b) What proportion of the scores lie within two standard deviations of the mean?
(c) What proportion of the scores lie within three standard deviations of the mean?

Exercise 2.1.12. For the data $323, 389, 332, 415, 289, 467, 407$ find the mean, variance and standard deviation.

Exercise 2.1.13. For the data $323, 389, 332, 415, 249, 467, 507$ find the mean, variance and standard deviation.

Exercise 2.1.14. The heights of the players on the NBA's Miami Heat's 2012–13 roster were

$$6'10'', 6'9'', 6'8'', 6'10'', 6'1'', 6'2'', 6'10'', 6'5'', 6'8'', 6'9'', 6'8'', 6'8'', 6'10'', 6'8'', 6'11'', 6'9'', 6'4''.$$

(a) What was the average height of a player on the roster?
(b) What was the median height of a player on the roster?
(c) What was the standard deviation of the heights?

Exercise 2.1.15. According to the U.S. Census Bureau, the number of motor vehicle accidents in the U.S. were as follows:

year	accidents (millions)
1990	11.5
1995	10.7
2000	13.4
2004	10.9
2005	10.7
2006	10.4
2007	10.6
2008	10.2
2009	10.8

(a) What was the mean number of accidents for these years?

(b) What was the standard deviation for these years?

Exercise 2.1.16. The SAT tests were intended to have a mean of 500 on both the reading/verbal and math scores. Here is some data of the mean SAT math scores in each of the years from 2003 to 2013:

year	2003	2004	2005	2006	2007	2008	2009	2010	2011	2012	2013
mean score	519	518	520	518	515	515	515	516	514	514	514

(a) What is the mean of these mean scores?

(b) What is the standard deviation?

(c) Do you think that the mean is 500 throughout the years? Explain.

Exercise 2.1.17. For the data in Table 2.2 on the home run leaders in Major League Baseball, the mean is 47.6 and the standard deviation is 8.02. If you drop out the data from the years 1998, 2000 and 2001, the mean drops to 45.64 and the standard deviation decreases to 4.73. Why is the standard deviation so much smaller when we drop out these three years?

Exercise 2.1.18. Suppose that a soccer team scores 77 goals and allows 63 goals in a 45-game season. How many games would you expect them to have won?

Exercise 2.1.19. Suppose that a soccer team won 36 of 45 games and allowed 51 goals in a 45-game season. How many goals would you estimate they scored?

Exercise 2.1.20. The following is the list of song durations (in the format minutes:seconds, so that 3:17 means 3 minutes and 17 seconds) for The Beatles' *A Hard Day's Night* album, which was released in 1964:

song title	duration (min:sec)
A Hard Day's Night	2:34
I Should Have Known Better	2:43
If I Fell	2:19
I'm Happy Just to Dance with You	1:56
And I Love Her	2:30
Tell Me Why	2:09
Can't Buy Me Love	2:12
Any Time at All	2:11
I'll Cry Instead	1:46
Things We Said Today	2:35
When I Get Home	2:17
You Can't Do That	2:35
I'll Be Back	2:24

(a) Convert all of the durations into seconds.

(b) What is the mean duration of a song on the album?

(c) What is the median duration of a song on the album?

(d) What is the standard deviation of a song's duration on the album?

Exercise 2.1.21. The following is the list of song durations for The Beatles' *Sgt. Pepper's Lonely Hearts Club Band* album, which was released in 1967:

song title	duration (min:sec)
Sgt. Pepper's Lonely Hearts Club Band	2:02
With a Little Help from My Friends	2:44
Lucy in the Sky with Diamonds	3:28
Getting Better	2:48
Fixing a Hole	2:36
She's Leaving Home	3:35
Being for the Benefit of Mr. Kite!	2:37
Within You Without You	5:04
When I'm Sixty-Four	2:37
Lovely Rita	2:42
Good Morning Good Morning	2:41
Sgt. Pepper's Lonely Hearts Club Band (Reprise)	1:19
A Day in the Life	5:39

(a) Convert all of the durations into seconds.

(b) What is the mean duration of a song on the album?

(c) What is the median duration of a song on the album?

(d) What is the standard deviation of a song's duration on the album?

(e) How do these compare with their earlier *A Hard Day's Night* album? What observations can you make?

Exercise 2.1.22. When is the standard deviation equal to 0? Why?

Exercise 2.1.23. Suppose you have a large data set and you have a new data value that is an outlier. Which of the following are likely to be affected by the outlier: the mean, median, mode, first quartile, third quartile, range, interquartile range, standard deviation? Explain your answers, giving an example where appropriate.

Exercise 2.1.24. Author Malcolm Gladwell has written a fascinating book, *Outliers*. In it he describes people who were so talented in their fields that they were far beyond the norm. Examples are people like Paul McCartney and John Lennon in songwriting, Bill Gates in technology, and Newton in mathematics and physics. Make a list of modern-day outliers and their fields of expertise.

2 | *Estimating Unknowns*

Sometimes we have all of the data available. For example, you might be able to see the test score of everyone in your class. You might find a website that has every at bat for every player on your favorite baseball team. You might know the incidence of heart disease in your extended family. But often, the entire group from which you want to gather data is enormous—much too big to actually gather the data from, or the gathering process might be too difficult. For example, perhaps you have a project where you need to know the income of everyone in the United States or the cancer rate among residents in a county. Time and money prevent you from doing what would really give you the truest picture—canvassing everyone, the whole **population**, everyone under consideration.

It is here that mathematics comes (yet again!) to the rescue. What you can do is pick a representative **sample**—a subset of the population—and take data from it. Provided this is done properly, statistics allows you to generalize your results from the sample to the whole population with high accuracy. Such predictions are called **point estimates**. An example will make things clear, I hope.

First, let's consider adult Americans' view on global warming—what percentage of people believe that global warming has already started? We could select some cities, scattered across the U.S., and from online phone books, select some people at random to contact and ask their opinion. Suppose out of 50 people contacted, 26 stated that they thought global warming had already begun. Can we extrapolate that to $26/50 = 0.52 = 52\%$ of all adult Americans feel this way?

Sampling from the Menu

Certain things seem clear from the example:

- Had we taken a smaller sample, we would expect the estimate to be less reliable.

- Had we taken a larger sample, we would expect that the point estimate would be closer to the true value (and had we contacted *every* person in question, we would have found the answer for certain!).

- If we contacted individuals completely at random, we would get the best estimate for the sample size.

So we have a number of issues to deal with. First, let's examine how best to choose a sample from a population to study. There are a number of general approaches that we can take:

- We can choose a sample purely at random—this process is called **simple random sampling**. In days of old, before computers, one used tables of random numbers, but nowadays, many programs have built-in procedure to generate random numbers, and one can use these, together with a list of the population, to choose the sample.

- We can break up the population into subgroups or *strata* (from the Latin word *stratum* for layer) and then choose within each group a random sample—a method called **stratified sampling**. The key to this is to make sure that those in a subgroup are *homogeneous*, that is, they are alike in some important way.

- We can choose a specific (small) number *k* and take every *k*th person from the population—this process is called **systematic sampling**.

- We can separate the population into groups or *clusters*, <u>not</u> based on any important characteristics, and then choose to sample everyone in one of the random clusters. This approach is called **cluster sampling**.

(Of course there are other, poorer methods, but they may suffer from serious biases. We'll talk more about this at the end of the chapter.)

To see the difference between these, let's imagine that we want to know what percentage of attendees at an event in a sports stadium have made a purchase, by the end of the game, at the concession stands. It is unwieldy to canvass all 44 000 fans in the sold out game, so we elect to choose a sample of 200.

- We could take a simple random sample by choosing 200 numbers at random from the ticket numbers that were sold and logged in as present.

- Realizing that 60% of the attendees are men, we could take a stratified sample, say by dividing the population (the attendees) into two groups, men and women. We then could choose say 120 men and 80 women at random.

- We could take a systematic sample by taking every 15th attendee entering the gates until we had a sample of 200.

- Each of the 220 sections has exactly 200 people, so we choose one of the 220 sections at random and sample *all* people in that section.

You probably feel that the simple sampling would provide the best data, and indeed it would; the reason the other methods are used in practice are

- simple random samples can be costly to gather (as they may involve, for example, traveling across an entire country or even the world to reach members of the sample) and

- simple random samples may be very hard to gather—one would need a list of all members of the population ahead of time, and those chosen for the sample would have to be available and willing to participate.

But while not ideal, the other methods have their advantages, too. The stratified method would allow the gathering of some additional data as well, such as, for example, how the sexes differ in their concession stand purchasing (if, by chance, in our simple random sample almost all were men, we couldn't do that). The systematic sample is pretty elementary to implement, but might pick too many people who bought their tickets early, and they might not be representative of the whole stadium population. Finally, the cluster sample sounds like a good idea, but if certain groups of people bought tickets together in the same sections (for example, the *Fasters'* Group or the *Foodies* Association) that might also skew the proportion of the sample who bought food at the concessions stands.

Now having sampled from a population, you would calculate some statistic for your sample, like a proportion, the mean, median, standard deviation, or some other number (such a value is called a **sample statistic** and varies from sample to sample, while the actual value for the whole population, the **population statistic**, doesn't vary at all). For example, suppose in the previous stadium example, you took a simple random sample of 200 attendees and found that the proportion of those who spent at the concession stands that day was 67%. Would you assume that was the proportion of *all* attendees who spent money at the concession stands? Of course not. You would think it might be pretty close to 67%, but would be shocked if it were exactly 67%.

So while our point estimate for the proportion of those who spent at the concession stands is 67%, we have to think that we need some wiggle room—saying that the proportion of *all* attendees who spent at the concession stands is 67% is almost surely incorrect. So we give what is called a **confidence interval** for the statistic, that is, we give a range, around the sample statistic, for which we have some level of confidence that the

You would also expect if you took some more random samples of 200 people that the proportion would vary a bit, perhaps up to 79%, perhaps down to 43%, and so on—perhaps even down to 0% if you happened to pick a group of dieters for the day. And it would certainly be unlikely that two samples would have *exactly* the same proportion!

population's statistic lies in. In our stadium example, it might be something like a range of 60.4% to 73.5%.

Definition 2.2.1. A **confidence interval estimate** for a population statistic, based on a sample statistic, is an interval, together with a **level of confidence**, which is the proportion of such intervals, if you repeatedly sample many times, that will contain the population statistic. The sample statistic is in the middle of the interval, and the **margin of error**, E, is the distance from the sample statistic to an end of the interval.

A confidence interval always looks like

$$[y - E, y + E]$$

where y is the sample statistic and E is the margin of error. Both y and E depend on the sample chosen, but E also depends on the confidence level as well (for a higher confidence level, E will be larger).

The definition is a little wordy, but it gets around the impreciseness of talking about how "confident" we are that the population statistic lies in the interval (though many people think of it in exactly this way!). So, for example, you might find that the 95% confidence interval was $[60.4\%, 73.5\%]$. That means that if you repeated this process (more about the details soon) many times, about 95% of the time, the population statistic (in this case, the actual proportion of the attendees who spent money at the concession stands) would be somewhere in the interval. The point estimate was 67%, smack dab in the center of the confidence interval, and the margin of error is $67\% - 60.4\% = 6.6\%$, or 0.066—not too much. That is the most you expect the error to be in your point estimate of the proportion 95% of the time.

It's unlikely that we would be completely confident that the range we pick would for sure contain the population statistic, unless it was HUGE. For example, we could take a range of 0% to 100%, but that would be ludicrously large and say nothing about the proportion. So there will be a necessary trade-off between the size of the interval and the confidence we have that it is correct—there is no free lunch, even for those who know statistics.

> *Life Lesson:* If you want to be more confident in the range for your estimate, you have to have a bigger range.

The actual calculation of the confidence interval depends, of course, on the statistic you are trying to estimate and can get a little involved. We'll look at just one application, namely, to proportions, as the math is a little easier. (The idea is much the same with estimating means of a population, or the differences between means between two populations, though the details are a little more involved, and we'll only pursue these in the exercises.)

Estimating Proportions

Let's begin with proportion. Suppose we want to estimate what proportion p of individuals in a population have a certain property—it might be a certain illness, a certain characteristic or a certain belief. To do so, we take a random sample of size n from the population and count the number $\hat{x}$ of individuals with that certain property. Our point estimate for the population proportion for the property is $\hat{p} = \hat{x}/n$. For example, a Gallup poll[1] taken March 6–9, 2014 of a random sample of 1048 adults found that 566 people believed that global warming has already begun. Based on this sample, the point estimate for the number of adult Americans who would believe that global warming has already begun is

$$\hat{p} = \frac{566}{1048} = 0.54,$$

that is, 54% of Americans believe that global warming has begun. But, of course, we don't really think that *exactly* 54% of Americans believe that global warming has begun, only close to this number. Suppose that our definition of *close* is "95% of the time," that is, our confidence level is 95%. Then our confidence interval for the proportion p of those who agree that global warming has

The little hats ˆ are to indicate that the quantity is based on a sample, rather than the whole population. Different samples will have different values (and possibly wear different hats!).

begun will look like

$$[0.54 - E, 0.54 + E]$$

where E is the margin of error (recall that the **interval** $[x, y]$ means the set of all numbers greater than or equal to x and less than or equal to y). How we calculate E for proportions is not too hard, though we'll leave out the explanation of how it is derived:

> Suppose we have taken a sample of size n from a population in order to estimate the proportion of the population with a certain property, and find that $\hat{x}$ in the sample have the property in question. The margin of error, for a confidence level of $c\%$, is calculated by
>
> $$E = z \times \sqrt{\frac{\hat{p}(1 - \hat{p})}{n}},$$
>
> where z depends on the desired confidence level. (The sample size n can't be too small for the calculation to work—you need $n\hat{p}(1 - \hat{p}) \geq 10$.)

The value of z comes from the standard normal distribution, being the value for which between $-z$ and z, $c\%$ of the values lie. The amazing fact is that, provided that the sample size is chosen appropriately, the sample proportion turns out to be normally distributed, with mean p (the true population mean) and standard deviation $\sqrt{\frac{p(1-p)}{n}}$ (for many of the population statistics you would want to estimate, the sample statistic turns out to be approximately normally distributed as well, if the sample is large enough). We saw in the previous section that 95% of the values for a normal distribution lie within 2 standard deviations—actually, we should have been a little more precise and said that within 1.96 standard deviations of the mean, 95% of the values lie. The number 1.96 is so close to 2, we rounded without mentioning it.

For our purposes, all we need is the following little table (which is small enough even to memorize, if you want!):

The definition of "close" in statistics took a while to be narrowed down. Most often, it means 95% of the time, though sometimes it is relaxed to 90% of the time or tightened to 99% of the time—it all depends on how often we are willing to be wrong!

If we sample with *replacement*, that is, we choose n items randomly from the population, returning the one we chose back into the population before we pick again, then the distribution is called a **binomial distribution**. If $\hat{x}$ is the number, out of the n chosen, that have the property in question (which has proportion p in the population), then the distribution of $\hat{x}$ is close to a normal distribution with mean pn and standard deviation $\sqrt{np(1-p)}$, and the distribution of $\hat{x}/n$ is close to a normal distribution with mean p and standard deviation $\sqrt{p(1-p)/n}$.

confidence level	90%	95%	99%
z value	1.64	1.96	2.57

So for our global warming example, we have found that $\hat{p} = 0.54$, and with sample size $n = 1048$, we find that our margin of error, at the confidence level of 95%, is

$$
\begin{aligned}
E &= z \times \sqrt{\frac{\hat{p}(1-\hat{p})}{n}} \\
&= 1.96 \times \sqrt{\frac{0.54(1-0.54)}{1048}} \\
&= 0.0301,
\end{aligned}
$$

or at most 4 percentage points (we round up, even though 3 would be closer, to make sure that our error is inclusive). That gives us a confidence interval of

$$[0.54 - 0.04, 0.54 + 0.04] = [0.50, 0.58].$$

That says that we wouldn't be surprised if the actual population proportion turned out to be any number between 0.50 and 0.58, but would be surprised if it turned out to be less than 0.50 or more than 0.58. If for many samples of size 1048 we created such an interval, we would expect that about 95% of the time, or **19 times out of 20**, the interval would contain the true population proportion of those who believe that global warming has started.

Now you know why, with the news' description of poll results, you always hear that the results are true "19 times out of 20."

If we wanted a higher confidence level, say 99%, we would find that

$$
\begin{aligned}
E &= z \times \sqrt{\frac{\hat{p}(1-\hat{p})}{n}} \\
&= 2.57 \times \sqrt{\frac{0.54(1-0.54)}{1048}} \\
&= 0.0395,
\end{aligned}
$$

again at most 4 percentage points, giving the same confidence interval of

$$[0.54 - 0.04, 0.54 + 0.04] = [0.50, 0.58].$$

We can state the same confidence interval, but with a higher level of confidence, so why don't we rewrite things? The point is that, for the sake of keeping our statistical noses "clean" and not rewriting based on the sample, we choose the confidence level *ahead of time* rather than afterwards.

So we've gone through the basic idea for point estimates and margins of error (and for finding confidence intervals) for a population proportion. But often in polls, there isn't just one proportion that is being estimated—there are actually a few of them. For example, the full poll on the question of global warming had a number of choices:

Already begun	54%
Within a few years	3%
Within your lifetime	8%
Not within your lifetime, but affect future	16%
Will never happen	18%
No opinion	2%

Now the margin of error, E, at the 95% confidence level (that is, 19 times out of 20) is given by

$$E = 1.96 \times \sqrt{\frac{\hat{p}(1-\hat{p})}{1048}}$$

as $z = 1.96$ from Table 2.4 and $n = 1048$ is the sample size. The issue is that $\hat{p}$, the sample proportion, changes from answer to answer. The sample proportions are $0.54, 0.03, 0.08, 0.16, 0.18, 0.02$, and plugging each of these into the formula above for E gives margins of error $0.0301, 0.0103, 0.0164, 0.0222, 0.0233, 0.0085$—all different! Rather than state a different margin of error for each answer, what is usually done is to simply state the **largest** of the margin of errors, as it will certainly encompass all of the others. Here we see that 0.0301, that is 3.01 percentage points, is the biggest of them, so rounding up, we can say that the margin of error is (at most) 4 percentage points.

Finally, picking the sample size and then calculating the margin of error is a little backwards from the way polling is done, simply because polling is expensive, and smaller samples cost less. So pollsters usually decide on an acceptable margin of error, say 4%, and choose the smallest sample size that will give (at most) this margin of error. So, for example, if we want a margin of error of 4 percentage points, we want $E \leq 0.04$, that is,

$$E = 1.96 \times \sqrt{\frac{\hat{p}(1 - \hat{p})}{n}} \leq 0.04,$$

and by a bit of simple math, we get

$$1.96^2 \times \frac{\hat{p}(1 - \hat{p})}{n} \leq 0.04^2$$
$$\frac{\hat{p}(1 - \hat{p})}{n} \leq \frac{0.0016}{3.8416}$$
$$n \geq 2401 \times \hat{p}(1 - \hat{p})$$

and now we appear to be stuck—we don't know what $\hat{p}$ is until we sample, and we can't sample until we know what the sample size n is!

But all is not as it appears—while we don't know what $\hat{p}$ is, we can say something about what the *biggest* $\hat{p}(1 - \hat{p})$ can be. Why? Elementary, my dear Watson, or should I say, high school? The reason is that $\hat{p}$ will certainly be between 0 and 1, and $\hat{p}(1 - \hat{p})$ can be expanded to look like $-\hat{p}^2 + \hat{p}$, which, if you replace $\hat{p}$ by x, I'm sure you recognize $-x^2 + x$ as a *quadratic that opens downwards*! Figure 2.11 shows a plot. It has its two x-intercepts at 0 and 1, and its vertex is halfway between them, at $1/2 = 0.5$, where it has a value of $-0.5^2 + 0.5 = 0.25$. This all means that while we don't have any idea of what $\hat{p}$ is, we do know for certain that $\hat{p}(1 - \hat{p}) \leq 0.25$, no matter what. And so,

$$2401 \times 0.25 \geq 2401 \times \hat{p}(1 - \hat{p}).$$

If we take n to be at least 2401×0.25, which is 600.25, we are sure to have the margin of error no more than 4 percentage points. So, as we are loathe to choose only parts of a person, we round up and know that a sample of at least 601 people will do.

It's not clear why the authors of the Gallup study chose $n = 1048$ instead of 601. I contacted them and it turns out that they didn't quite do a simple random sample—their process involved a more complicated sampling procedure, and that is the reason that the sample is larger than it would have been otherwise. In any event, having more in the sample won't hurt!

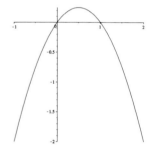

Figure 2.11: The quadratic $x(1 - x) = -x^2 + x$. Isn't it wonderful how high school mathematics revisits you in statistics?

Confidence Intervals and Experimentation

Researchers, while interested in confidence intervals, often turn the idea on its head. They phrase their research as an experiment that aims to investigate a statement or claim. For example, suppose there was the claim that

H_0: The proportion of adults in the U.S. who think that global warming is occurring is $p = 0.51$, that is 51%.

This original claim is called the **null hypothesis**, sort of a baseline hypothesis; the opposite hypothesis, called the **alternate hypothesis**, H_{alt}, is also put forward, as an alternative what might be true:

H_{alt}: The proportion of adults in the U.S. who think that global warming is occurring is <u>not</u> 0.51, that is, $p \neq 0.51$.

These hypotheses give rise to a **two-tailed test**, so-called as we reject the null hypothesis if the sample statistic is either too small *or* too high—that is, if it is in either of the two extremes.

Given that we have found that 95% of the time the actual population proportion (of those who believe that global warming has begun) is in the range $[0.50, 0.58]$, and 0.51 is in this range, we would have to say that there isn't enough evidence, based on the sample, to suggest that the population proportion isn't 51%, that is, **there isn't enough evidence to reject the null hypothesis**. It's a rather weak statement, saying we can't say something more definitive (and perhaps more interesting), but researchers are a conservative lot—they need a fair bit of evidence to make a strong claim (like the alternative hypothesis).

On the other hand, if the null hypothesis was that

H_0: The proportion of adults in the U.S. who think that global warming is occurring is 0.40, that is, 40%.

then, as 0.40 is well outside the 95% confidence interval $[0.50, 0.58]$, we see this as highly unusual, picking a sample like the one chosen less than 5% of the time, if the proportion were truly 0.40. Thus in this case we would **reject the null hypothesis** and accept the alternative hypothesis, which would be

H_{alt}: The proportion of adults in the U.S. who think that global warming is occurring is <u>not</u> 0.40.

So providing confidence intervals or accepting or rejecting hypotheses are closely related, with the choice of which to use depending on what kind of conclusion you are after—a range of values or the validity of a claim. Of course, there are times when the claim that you are interested in isn't that a population statistic is different from a particular value, but that it is *greater than* or *less than* a particular value. For example, if the incidence rate for diabetes is known to be 7.8% of the general adult population, a research group might investigate the validity of the statement

vigorous exercising for one hour each day decreases the rate of diabetes.

Mathematically, we might state the null hypothesis (the status quo) as

H_0: The proportion of adults who exercise vigorously for one hour each day and have diabetes is 7.8% or more.

The more interesting alternative is

H_{alt}: The proportion of adults who exercise vigorously for one hour each day and have diabetes is less than 7.8%.

We would then conduct an experiment by randomly selecting a sample, and then subject them to an hour of vigorous exercise each day, for a significant time period. For such hypotheses, we only reject the null hypothesis if the sample statistic is too extreme in one direction, but not if it is extreme in the other; in our diabetes example, we would only reject the null hypothesis (that the proportion of adults who exercise vigorously for one hour each day and have diabetes is 7.8% or more) if we found that the sample proportion was significantly less than 7.8% (if it turned out to be significantly more, this would <u>not</u> provide any evidence to reject the null hypothesis). In this case, these hypotheses give rise to a **one-tailed test**—we will reject the null hypothesis only if the sample statistic is too extreme in one direction (but not in the other).

In such a one-tailed test, confidence intervals are not used. In fact, for both one- and two-tailed tests, what is commonly done is some statistical work that provides what is called a *p-***value**, which is the probability that you would get a sample statistic this extreme or more extreme, just by chance, if the null hypothesis holds. The details are more involved than we want to go into here. But what is important is the idea—a small p-value is a good indication to reject the null hypothesis. How small is small? The smaller the better!

To err on the side of caution, we want to limit the probability of a **Type I error**, that of rejecting the null hypothesis when it is, in fact, true—this may happen, but it should be rare (the probability of a Type I error is called the **level of significance**). How rare? The usual standard is what we have seen—only 5% of the time (though when we are even more wary of the implications of rejecting the null hypothesis, we might take "rare" to mean at most 1% of the time). Having a calculated p-value of 0.0395 would give good evidence that the null hypothesis should be rejected (that is, *we reject the null hypothesis at the* 0.05 *level of significance*), and we should embrace the alternate hypothesis, and we should do so with even more vigor if the p-value were smaller, say 0.00212, as getting the sample statistic we do get is even more unlikely, were the null hypothesis to be true. On the other hand, if our p-value were 0.254, this would be above our threshold of 0.05 (or in more extreme cases, 0.01) for rarity, and we would only be able to say that we couldn't reject the null hypothesis. It is really important to realize that this is <u>not</u> the same as saying we state that the null hypothesis is true—we only can commit to saying we don't have enough evidence to say it is false. It may well be false, but we don't have enough to go on to go out on a limb and say so.

Unlike the diabetes example, we often don't know the population statistic. Then the gold standard for proving a cause and effect (and really the only convincing one to scientists and statisticians) is to carry out a **randomized control trial**, where a large random sample from the population is broken into two groups randomly—one receives some new treatment, the other (the **control group**) does not—and we compare a given sample statistic, such as proportion or mean, for both samples. As long as the samples are large (usually this means at least 30),

• the difference between proportions approximately follows a normal distribution, and

The choice of the use of "p" in "p-value" is rather unfortunate, as we have used p before to stand for a population proportion, but it is standard, and there are only so many English letters to go around!

- the difference in means approximately follows what is called a **Student's *t*-distribution**, which is very much like a normal curve (only "fatter" at the ends, depending on the sample sizes).

Sometimes, it is impractical or unreasonable (or even unethical) to carry out a randomized control trial (imagine subjecting some people to smoking cigarettes to see how many get cancer!), so an **observational study** is a often done, where some participants have chosen previously to subject themselves to a certain treatment, others not, but the results of such studies are much less convincing about causation.

All sorts of interesting, unusual and thought-provoking questions can be analyzed with hypothesis tests. For example, one that just came across my desk is called *Eyes in the Aisles: Why is Cap'n Crunch Looking Down at My Child?*, a study of whether the characters on children's cereal boxes make eye contact with children walking down store aisles, and whether this influences the choice of brand. Once they classified cereals as being advertised to parents (to give to their children) or directly to the children themselves, the authors of the study determined mathematically the height that the cereal boxes "spokes-character" appeared to be gazing at a distance of 4 feet away, where someone walking down the aisle would likely be (we'll talk in a later chapter about how this calculation was done). They also used Photoshop to change the eye gaze on packaging from looking at the viewer to looking away (the latter being the **control group**, those for which the "treatment" was omitted). They found a few interesting things:

- The average eye angle of gaze for children-directed cereal spokes-characters was down $9.67°$, versus up $0.43°$ for adults, and this was highly significant ($p < 0.0001$).

- The average height, at 4 feet away, of the characters' gazes for boxes marketed to kids was 20.21 inches for children and 53.99 inches for those aimed at adults, again very significant ($p < 0.0001$).

- For those viewing boxes where the spokes-characters were making contact with them, viewers reported that they felt a greater connection to the cereal with a p-value of 0.05, and feelings of trust with the cereal were just a bit higher, $p = 0.65$.

Trix® aren't just for kids—they can be played by spokes-characters on any unsuspecting customer! The authors drew some conclusions from the experiments and even made some recommendations based on them (for example, eye contact with spokes-characters could be used to market healthier cereals to children).

Finally, there is a **Type II error** that is present in any experiment, namely, the probability of <u>not</u> rejecting the null hypothesis when it is in fact false. The Type II error is harder to calculate, as any calculation for it likely depends on exactly what the unknown population statistic really is. But the Type II error is nonetheless always lurking about, and it is a balancing act—if you try to decrease the Type I error, you inevitably increase the Type II error, and vice versa. So how do you set the scales? You are ultimately guided by the consequences of each of the errors occurring—how bad is it for you to reject the null hypothesis when it is true, versus not rejecting it when it is false?

Exercises

For Exercises 2.2.1 to 2.2.6, decide what kind of sampling (simple random, stratified, systematic or cluster) is described.

Exercise 2.2.1. From a school population of 2000, the students are arranged in a list, and 35 students are chosen randomly.

Exercise 2.2.2. From a school population of 2000, an entire class is chosen randomly, and all students are included in the sample.

Exercise 2.2.3. From a school population of 2000, the students are grouped by age, and 20 students are chosen randomly from each age group.

Exercise 2.2.4. From a school population of 2000, every 50th student leaving the building is included in the sample.

Exercise 2.2.5. From the collection of this year's SAT tests, the collection is divided up by each state, and all tests from three randomly chosen states are selected.

Exercise 2.2.6. From all cars sold in North America, the names of everyone who has purchased a car is entered into a computer and 375 are chosen at random.

For Exercises 2.2.7 to 2.2.10, decide whether the statistic mentioned is a population or sample statistic.

Exercise 2.2.7. The number of students in a school who have had breakfast.

Exercise 2.2.8. Among a randomly chosen class in a school, the number of students who have had breakfast.

Exercise 2.2.9. The proportion of Americans who have visited Europe.

Exercise 2.2.10. Among every 15th visitor at a specific sporting event, the proportion who have visited Europe.

Exercise 2.2.11. Suppose we have found that in a random sample of 50 American adults (18 years or older), 11 of them smoke.
(a) What would be your point estimate for the proportion p of adults in the U.S. who smoke?
(b) Calculate a 90% confidence interval for p. What is the margin of error?
(c) Calculate a 95% confidence interval for p. What is the margin of error?
(d) Calculate a 99% confidence interval for p. What is the margin of error?

Exercise 2.2.12. Based on your answers to Exercise 2.2.11, at what level(s) of confidence would you believe that the true proportion of smokers in the U.S. was 18.1%?

Exercise 2.2.13. A Gallup poll (March 6–9, 2014) of 1048 American adults (18 years or older), asked them "Next, thinking about global warming, how well do you feel you understand this issue?" with the following responses (Copyright ©2014, Gallup Inc. All rights reserved. The content is used with permission; however, Gallup retains all rights of republication.):

Very well	33%
Fairly well	51%
Not very well	14%
Not at all	2%
No opinion	0%

If the margin of error is 4% at the 95% confidence level, find confidence intervals for each proportion.

Exercise 2.2.14. Suppose a poll of a proportion wants to have a margin of error of 2%. If the confidence level is 95%, what sample size is needed?

Exercise 2.2.15. Repeat Exercise 2.2.14 if the confidence level is 90%.

Exercise 2.2.16. Repeat Exercise 2.2.14 if the confidence level is 99%.

Exercise 2.2.17. Suppose that a poll samples 1500 people in order to estimate a population proportion.

(a) If the desired confidence level is 90%, how big can the margin of error be?

(b) If the desired confidence level is 95%, how big can the margin of error be?

(c) If the desired confidence level is 99%, how big can the margin of error be?

Exercise 2.2.18. Suppose an experiment for the effect of a new antiviral drug has been carried out. The null hypothesis is that the drug works no better than a placebo. What can you say in each of the following cases?

(a) The *p*-value turns out to be 0.230.

(b) The *p*-value turns out to be 0.066.

(c) The *p*-value turns out to be 0.021.

(d) The *p*-value turns out to be less than 0.001.

Exercise 2.2.19. Take 10 coins. Shake them all together in your hands and toss them in the air. Count the number of heads. Repeat this 30 times; this yields a binomial distribution with $p = 1/2, n = 30$. Make a frequency plot of the number of heads tossed. What does the curve look like?

Exercise 2.2.20. Repeat Exercise 2.2.19 but instead of 10 coins use 10 dice, counting the number of 6's in each toss.

Finding confidence intervals for means: How one deals with confidence intervals and hypothesis tests for means and for the difference between proportions involves distributions different from the normal distribution. As mentioned earlier, they are called the **Student's *t*-distributions**, each with its own number of **degrees of freedom**, which is a positive integer based on the sample size. These distributions look much like the normal curve, but are fatter in the tails; as the degrees of freedom gets large, the curve approaches the normal curse. Figure 2.12 shows the normal curve, along with Student's *t*-distributions with 5 and 10 degrees of freedom; note how the tails are a bit fatter than the normal curve, but the Student's *t* curves get close to the normal curve as the degrees of freedom grows (for a large number of degrees of freedom, there is practically no difference between the Student's *t*-distribution and the normal distribution).

There are tables for looking up values in Student's *t*-distributions, and computer programs that will do so as well. Here are some *t* values (a *t* value for a confidence level of $\alpha\%$ means that $\alpha\%$ of the data lies between $-t$ and t in the Student's *t*-distribution):

For 5 degrees of freedom:

confidence level	90%	95%	99%
t value	2.015	2.571	4.032

For 10 degrees of freedom:

confidence level	90%	95%	99%
t value	1.812	2.228	3.169

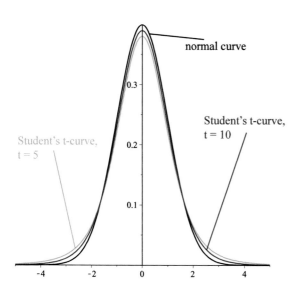

For 50 degrees of freedom:

confidence level	90%	95%	99%
t value	1.676	2.009	2.678

For 1000 degrees of freedom:

confidence level	90%	95%	99%
t value	1.646	1.962	2.581

Note how close the last table is to the corresponding table for the normal distribution.

Suppose that the data is normally distributed or the sample size n is large. Our point estimate for the population mean is the sample mean $\bar{x}$, and the margin of error (for a given confidence level) is

$$E = t \times \frac{s}{\sqrt{n}}$$

where t is the corresponding t-value with $n - 1$ degrees of freedom, and s is the **sample standard deviation**, which is calculated just like the population standard deviation, except you divide by $n - 1$ under the square root sign instead of by n.

Exercise 2.2.21. Heights of adult females are normally distributed. If you take a simple random sample of size 6 from the population of all American adult women, and get the following: $5'4'', 5'9'', 5'3'', 5'6'', 4'11'', 5'7'',$

(a) Find the sample mean and sample standard deviation.

(b) For a confidence level of 95%, find the margin of error and the confidence interval for the mean height of adult women in the U.S..

Exercise 2.2.22. Weights of adult males are normally distributed. If you take a simple random sample of size 11 from the population of all American adult men, and get the following (in pounds): 184, 177, 198, 191, 149, 210, 207, 183, 188, 179, 199,

(a) Find the sample mean and sample standard deviation.

(b) For a confidence level of 99%, find the margin of error and the confidence interval for the mean weight of adult men in the U.S..

Exercise 2.2.23. Suppose with the stadium example, we select 51 attendees and question them after the game about the *amount* they spent at the concession stands that day. Suppose that you find that the average amount spent in the sample was $\bar{x} = \$53.57$, with a sample standard deviation of $s = \$14.02$. Assume that the amount spent is normally distributed.

(a) Find a 90% confidence interval for the average amount spent by attendees at the concession stands.

(b) Find a 95% confidence interval for the average amount spent by attendees at the concession stands.

(c) Find a 99% confidence interval for the average amount spent by attendees at the concession stands.

Exercise 2.2.24. For each part in Exercise 2.2.23, what is the margin of error?

Exercise 2.2.25. For each part in Exercise 2.2.23, what would you say to the belief that the actual average amount spent at the concession stands was $58.05?

Exercise 2.2.26. For each part in Exercise 2.2.23, what would you say to the belief that the actual average amount spent at the concession stands was $54.45?

Exercise 2.2.27. If the sample size is 1001 and the margin of error is 4.21 for the confidence interval at a confidence level of 90% for the mean, what is the sample standard deviation?

3   *Leading You Down the Garden Path with Statistics*

Statistics are a great thing—they provide you with numbers that can encapsulate huge amounts of data, making the latter palatable and understandable. But, in the wrong hands, they can confuse and even mislead, either accidentally or intentionally. Writer Mark Twain wrote in 1906,

> *"Figures often beguile me, particularly when I have the arranging of them myself; in which case the remark attributed to Disraeli would often apply with justice and force: 'There are three kinds of lies: lies, damned lies, and statistics."*

The only thing you can truly rely on to protect you is the gray matter between your ears. So let's explore a few instances where statistics can lead you astray, "down the garden path" so to speak, in order to learn what (and who) to look for.

Garden Path #1: Sampling Bias

We've seen that when someone wants to estimate a population statistic, the reasonable thing to do is to choose a sample and estimate the population statistic from a the corresponding sample statistic. Certainly there can be problems if the sample size is too small. For example, if you wanted to determine the rate of cancer in your state and only sampled 5 people at random, you could find no instances of cancer, and predict a cancer rate of 0%, or find 3 out of the 5 having cancer and predict a rate of 60%, neither being likely the true answer. Sampling more is always better, and often crucial to the accuracy and predictability of the results. After all, if you only took a sample of size one, you would predict either 0% or 100% of people had cancer!

But even if you choose a large sample, it is vital that the sample be chosen in a way that is as *representative* as possible of the entire population. The gold standard is to choose a simple random sample, where all of the members of the population are equally likely to be chosen. Of course, sometimes that may be difficult or even impossible, so other approaches, such as stratified, systematic or cluster sampling are used, but the further you get from simple random sampling, the more likely you are to introduce **bias** into the sample, and that may make the estimation less accurate. For example, if you took a poll of your friends and family on global warming you might end up with a markedly different result from the Gallup poll described in the previous section. I don't know your views on the subject, but chances are that those closest to you share some of your views, and taking even a large sample won't help if the sampling is far from random. Even if you veer far away from friends and family and call random cell phone numbers from an online list for your city, there is at least one issue:

- You only reach those who have a cell phone, and leave out those who only have land lines or those who have no phones at all! And you've only reached a sample in your city. The demographic of people you've reached may be markedly different from the general population. (non-representative sample)

- You may only reach certain people when you call—what about those who don't answer their phone—are they likely to have the same views as those who do? Many times pollsters are unable to reach certain individuals in their sample, and again, there may be marked differences in characteristics and views between those who are reached and willing to participate and those who can't be reached or who are unwilling to participate. (non-respondent bias)

Any of these make any statistical calculations, even those done perfectly correctly, suspect. And biases can creep in, as they often do, quietly, in other ways:

- Participants may choose themselves to be part of the sample. For example, internet polls are often this way—but those who choose to participate may be far different from those who don't. Also, participants in a sample may be drawn from those who volunteer or are willing to accept payment for participation. (self-selection bias)

- People may drop out of a study due to various reasons, perhaps some reflecting the study itself, and omitting their data may bias the results. (dropout bias)

And the list goes on. One well-known example of sampling bias occurred in the 1936 U.S. elections. Prior to the election, the magazine *Literary Digest* sent out over 10 million surveys in the mail, receiving back roughly 2.3 million, and based on these, the magazine predicted that incumbent Democratic president Franklin Roosevelt would go down to defeat to his Republican opponent, Alf Landon, whom they forecast would win about 60% of the popular vote. Remember Alf Landon? I'm pretty sure you don't, as he lost handily to Roosevelt and wasn't heard of politically again.

What was the problem? Certainly the sample was large, but that didn't reduce the problem that the sample was biased.

The respondents were selected from readers of the magazine, registered voters, and owners of automobiles and telephones (these were all easily reachable by the magazine). These people tended to be wealthy—and Republican, and ultimately were not representative of the general voting population. Also, there was a problem with a very low response rate to the survey—only about 23% bothered to mail back the surveys. Who would likely send them in? Mostly those who were unhappy with the current president, and that skewed the results (badly!) as well. The non-representative, non-respondent and self-selection biases reared their ugly heads all at once, evidence of the fact:

Life Lesson: When things go south badly, it is often more than one problem at fault.

It's interesting to note that there was an up and coming pollster named George Gallup who predicted accurately that not only would Roosevelt win, but that the *Literary Digest*'s poll was incorrect. He based his predictions on a survey of a mere 50 000 randomly selected individuals, from the same lists as the *Literary Digest*'s survey. Gallup polls have been a gold standard in polling since then.

But that is not to say that even Gallup polls are always right. In 1948 a Gallup poll erroneously predicted that Democrat president Harry Truman would be defeated by Republican Thomas Dewey. On the day after the vote, Truman gloated in a now famous photograph by holding up a copy of the *Chicago Tribune* newspaper that declared, "DEWEY DEFEATS TRUMAN." Two reasons conspired in the embarrassingly wrong poll results. First, the Gallup poll relied on **quota sampling**, where specific numbers of individuals were chosen to be polled in various categories, by gender, race, income, etc. However, interviewers were given latitude in selecting those in each category, as long as they met their "quotas," and this was a source of bias. Overall, the quota sampling was less representative than simple random sampling. But perhaps more importantly, the polls was taken

Figure 2.13: Dewey Defeats Truman?

weeks before the election, and what might have been a more accurate statistical snapshot then was out of focus at election day. Sometimes, indeed, timing is everything.

Garden Path #2: Interviewing Bias

Sometimes the issue is not with the sampling but with the questions being asked. One has to be very careful when polling not to ask a question that encourages a particular response. For example, if you were asking about views on concussions in football, what type of response do you think you would get if you asked "Do you think that the National Football League is being negligent in protecting players from concussions?" as opposed to "Do you think that concussions are an issue for players of the National Football League?" I would think that the former almost begs the survey participant to say "yes." Restating the question as "Football is a tough sport. Do you think that football players are overly worried about concussions?" would likely elicit more "no" responses. Likewise, asking Americans whether they would like their government intervening somewhere in the world to support democratic values is likely to get a different response if the poll asks would they like their government intervening somewhere in the world by sending troops into harm's way, to support democratic values.

Many medical statistical studies are done **double-blind** to keep both the patient and the experimenter unaware of who is getting the new treatment and who receives the **placebo**, the simulated treatment that had no effect, as (i) patients who know they are getting the treatment may respond differently than they would otherwise, and (ii) the experimenters who know which patients are getting a new treatment may unconsciously (or consciously) record the results differently than they would otherwise. There was a famous study (*Newscasters' Facial Expressions and Voting Behavior of Viewers: Can a Smile Elect a President?*, B. Mullen et. al., J. Personality and Social Psychology 51 (1986), 291–295) where the effect of newscasters' facial expressions on voters was investigated. When comparing three famous newscasters for their positive or negative facial expressions while reporting on the Reagan–Mondale 1984 presidential race, the

Sometimes there are problems with the interviewers themselves. Interviewers are often given quotas for how many people to interview from a variety of categories (such as race, income level, gender, etc.) but are given leeway as to who precisely to interview. So interviewers sometimes avoid certain households or certain areas of a city, and sometimes they choose those who are most readily available at home during the day, which may have in higher proportions mothers and the unemployed—meaning that the sample is likely not representative of the whole population. There have even been cases of interviewers for major polls making up responses for imaginary respondents!

authors of the study found significantly more positive expressions, such as smiling (though no more positive reporting), from one of the news anchors, ABC's Peter Jennings, toward Ronald Reagan, than from the others (CBS and NBC), and it was noted that significantly more viewers of ABC planned to vote for Reagan. So if something as subtle as facial expressions may change voters' minds, the art of interviewing needs to be scrutinized very carefully before putting forward any poll results.

Also, there can be interviewer bias when the interviewees don't understand the questions! For example, when the major pollsters were interviewing people about possibly impeaching President Richard Nixon back in the 1970s, over 50% of Americans didn't understand what precisely "impeachment" entailed. But that didn't stop the pollsters from asking whether poll participants were in favor of Nixon being "impeached and compelled to leave the presidency" (impeachment only entails being charged with an offense and being tried by the Senate—not being found guilty and sentenced too!). Interviewers still pushed for answers, which they often do, even when the interviewees may not understand the question or are ambivalent about their views.

Garden Path #3: People Sometimes Lie

Sometimes we can do everything right in polling, but it's difficult to get an honest response. People will likely tell the truth about what they ate for lunch (maybe), but try asking them about their income, their truthfulness on tax forms, and their sexual preferences—they are likely to lie, or at least give a guarded response. And people even lie about whether they will vote or not (to say they won't may seem unpatriotic). Even the best worded questions may elicit suspect answers.

One answer to such delicate questionnaires comes from a very clever idea—chance! Suppose we have a yes/no question for which we have reason to believe some people are inclined to lie, such as "Have you stolen office supplies?" What can be done is after asking the question, to ask the participant to flip a coin. If it comes up heads, the participant is to respond truthfully, and if it comes up tails, he or she is just to respond "no." What is the end result? The participant feels secure, as even the interviewer can't know whether the response is truthful, so there is no reason to lie. Does that mean that the results are unusable? Far from it. Suppose from a large sample, you find that 64% said "yes" to the question. We'd expect about 50% of the yes answers came from those who flipped tails, as there will be about 50% who do flip tails. That leaves 64% − 50% = 14% of the other approximately 50% who flip heads, and answered truthfully, so our best estimate for the percentage of truthful "yes" answers would be 14% out of the 50%, or 14/50 = 0.28 = 28%. Isn't that brilliant?

Garden Path #4: Are You Measuring the Right Thing?

A statistic can be completely correct, but can mislead, as it may not be the relevant statistic to look at. For example, suppose you are interested in income for Nashville musicians. It may turn out that the *mean* income is quite high, say perhaps hundreds of thousands of dollars, as there are a few Nashville songwriting and performing cats who earn a lot of money. But these are outliers, and we have seen that outliers can affect the mean greatly. So what are the relevant statistics? I think the median would be a better score to look at, and on this account, it might very well turn out that the median income is much smaller, perhaps only $20 000, as most Nashville musicians are probably tending bars and waiting tables rather than gracing the stage of *The Grand Ole Opry*.

You might keep the issue of mean versus median in your mind when you go for job interviews—a high salary for a CEO and other top administrators will probably raise the mean, while the median salary may be a better number for you to go by.

On the other hand, sometimes it can be the mean that is more important. Suppose you are wondering whether to undergo an experimental drug treatment for a fatal disease. You might find out that the median life expectancy on the treatment plan is almost the same as that without and decide to refuse it. But what if the mean life expectancy with the treatment plan changes from 2 years to 5 years, as those who respond to treatment are essentially cured and live a long life? That is significant and shows that the median is not always the most important statistic.

Moreover, maybe neither is for some issues! If you are an investor in the stock market, you may value a lower average return over the years that has lower volatility, that is, lower spread, than one with a higher average gain but more of a roller coaster ride, with big ups and downs. You personally hate that and want to avoid it. You just aren't that big of a risk taker. Then a statistic of great importance will be the standard deviation, and just the mean or median won't do for you.

You also need to be careful when reading through articles as to whether the quoted statistical results refer actually to the appropriate population. For example, I heard about someone who came down with multiple sclerosis (MS) after a head injury. I went and looked up studies about the association, and there were quite a few studies done, with most showing no statistical connection between having sustained a head trauma and subsequently developing MS. It seemed like the road would end there, but the person who contracted MS after hitting her head was in her late twenties, and the study I had looked at was for both children and adults. A more specific study for those who experienced head traumas <u>after</u> age 20 did show an statistically significant connection—this connection was likely hidden when people of all ages were included in the first study. But the population most like the individual I had heard about was more specific than that, and hence a comparison to the appropriate adult population was much more relevant.

As in the last example, statistics can be used to find relations between two different quantities—are higher SAT scores associated with owning expensive computers? Statistics can be used to find the **correlation** between two quantities, a measure of how connected they are. Correlation is always between −1 and 1, with

- a large positive value meaning that when one quantity increases, so does the other,

- a large negative value meaning that when one quantity increases, the other decreases, and

- a value near 0 means that the increasing or decreasing of one quantity has no effect on the increasing or decreasing of the other.

We'll talk a bit more about correlation in the next chapter. The calculation is a little complicated, but similar to that of calculating the standard deviation. The key thing is that finding that two quantities are highly correlated does not imply that one causes the other. The mantra to remember is

Life Lesson: Correlation is not causation.

So one would likely find that high SAT scores are indeed positively correlated with owning expensive computers, but it would be a mistake to think that owning expensive computers *caused* higher SAT scores, or even that higher SAT scores caused one to own expensive computers. What is more likely, in this case, is that some other factor, like wealth, is a cause of both—wealth allows one to buy pricey items as well as to higher SAT tutors. But the temptation is to think, whenever a strong correlation is drawn, that one factor must be causing the other.

Garden Path #5: Statistics Aren't Always *Right*

Most people assume that all polls are correct—always. But we've seen that polls have a trade-off between accuracy and confidence level. They are likely to be correct, within the stated margin of error, 19 times out of 20 (at the 95% confidence level), but that yields the following, perhaps, surprising result:

Life Lesson: For a large number of polls, you can expect that about one in twenty will be incorrect!

That's right—the error is built right into the process—it can't be avoided. So you have to, have to, have to be wary of poll results. Sometimes they just have to be wrong—you just don't know when. This has broad applications to research science as well—for every 20 experiments carried out, you can expect one quoted result to be wrong. The only way to mitigate the error is to repeat your experiments more than once—that will drive down the chances of seeing the same erroneous result (but not remove it).

Research institutions and journals, whose reputations are on the line for the research that they publish (and thereby give their stamp of approval) are aware of such problems. An unscrupulous research scientist could repeat an experiment many times over, looking for a significant result. Perhaps none exists, but just by the "19 times out of 20" litmus test, if he repeats the experiment enough, one of the times he'll get an extreme result, extreme enough to draw the conclusion he wants. And at that point, he can ignore the many, many times that no significance was found and try to publish the one experiment that did provide the statistical evidence he or she desired. The statistics are not incorrect, but the interpretation is what is at stake, from all of the "non-results" that are swept under the scientific rug. In fact, many institutions and journals require those who carry out research to register their experiments with them *prior* to beginning them, so they can track the number of such tests that don't lead to significant results.

Garden Path #6: Statistics Aren't Always *Important*

Sometimes all of the sampling, questioning and statistics are correct, and you arrive at a statistically significant result. For example, you might find that there is indeed a difference between the average salary for those 5 or more years out of college for those who have taken an English course as opposed to those who haven't. Sounds positive, right? But what if the mean salary for those who have taken English in college is \$65 700, while for those who haven't it is \$65 800? With a large enough sample, this result could certainly be significant, statistically, but is it important? Does the extra \$100 make a difference to those who are earning? Probably not. So any statistically significant result still has to satisfy the true litmus test—is it important enough?

And suppose that you find out that for a pre-election poll 57% of respondents say that they will vote for Candidate A while only 43% support Candidate B. But because the sample was small (say perhaps 20 people), the margin of error is 8

percentage points. Should you declare A the winner? That would be too hasty, because within the margin of error, A's support could be as little as $57\% - 8\% = 49\%$, with B's support as high as $43\% + 8\% = 51\%$, so B could indeed win with the poll still being correct! Now, of course, A's support could be $57\% + 8\% = 65\%$ and B's as low as $43\% - 8\% = 35\%$, which would be a landslide victory for A, but we can't be sure. So as much as we'd like to declare a winner, our safest call is that the election is too close to call, within the margin of error.

Exercises

For Exercises 2.3.1 to 2.3.14, describe the type of bias that has entered or may have entered the statistical process.

Exercise 2.3.1. The sample of people chosen for a weight study was taken as every 5th person entering a fast food court in a mall on a certain day.

Exercise 2.3.2. You poll visitors to your website as to their views on abortion.

Exercise 2.3.3. You randomly select people from your university population to ask them about whether they have plagiarized an assignment, in hopes of estimating how many students at your university have indeed copied assignment work.

Exercise 2.3.4. An interviewer has a list of selected households to poll, but avoids ones where there is a dog out front.

Exercise 2.3.5. You advertise for individuals to take part in a psychology study on the effects of video gaming on reading.

Exercise 2.3.6. You take the average grade in a class that is **bimodal**, that is, the scores are clumped at both ends—everyone either gets a very low score or a very high score.

Exercise 2.3.7. A researcher has submitted to a top journal a purportedly groundbreaking article in which strong statistical evidence is shown to correlate listening to music with improved motor skills.

Exercise 2.3.8. An interviewer calls a randomly selected household to poll about the upcoming presidential election and polls everyone in the household: the father, mother, 24-year-old-son and 17-year-old daughter.

Exercise 2.3.9. An interviewer has a quota to reach of 60 males and 40 females from the U.S. adult population and goes to a local bar to canvas males, and a local gym to canvas females.

Exercise 2.3.10. The interviewer is to ask members of the selected sample about gun control. She asks, "Do you feel that background checks are necessary, or do you feel that they go against the freedom that every American loves?"

Exercise 2.3.11. The median time for waiting at airport security is found to be quite low.

Exercise 2.3.12. The mean wait time for knee replacement surgery is found to have dropped dramatically.

Exercise 2.3.13. A sample of those who have had an adverse drug interaction between two specific drugs is chosen from a group of litigants in a trial against the drug companies.

Exercise 2.3.14. A random sample shows that there is a high correlation between car accidents and having a Facebook account, and hence one should delete their Facebook account in order to lower their risk of a serious driving accident.

Exercise 2.3.15. In Exercise 2.3.14, is it *impossible* that having a Facebook account might lead to car accidents? How might you check this out statistically?

Exercise 2.3.16. Suppose you read about an experiment where it is claimed that the p-value is 0.04, and hence the null hypothesis should be rejected. What might be the problem with this reasoning?

Exercise 2.3.17. The authors of the study into the effect of newscasters' facial expressions in swaying voters found that their null hypothesis that the proportion of ABC viewers voting for Reagan was the same as that for NBC and CBS had a p-value of 0.0053. How statistically significant is the result?

Exercise 2.3.18. There is a statistical principle called **regression to the mean** which states that if, over time, the mean and variation don't change for some distribution, then extreme values tend to be followed by less extreme ones. This is somewhat paradoxical, as we like to think of trends. For example, the children of top hitting baseball stars might turn out to be baseball players, but are most often not quite as good as their parent. How would this principle inform your decisions as a manager of a baseball team?

Exercise 2.3.19. Make a list of people who were extreme in their field, whether it be science, mathematics, art, music, and so on. Then try to think of those whose children were even more extreme! What do you notice?

Exercise 2.3.20. Stein's paradox in statistics says that short term averages of individuals are not the best indicators of long term averages—they inform them to some extent, but it needs to be adjusted by the averages of all those in the population. So, for example, a baseball player who is on a hitting streak (say batting 0.345) during the first half of a season is likely to be batting well over the rest of the season, though not as well as in the first half, as the average batting average of all players is (which is around 0.250). Does this have implications for general managers as well?

And Now the Rest of the Story ...

And now we return to Dr. Wakefield's research on the connection between vaccinations and autism. His research paper was based on twelve children with autism-like symptoms who were referred to London's Royal Free Hospital. Of these, eight of them had parents or doctors who noted that the developmental symptoms started soon after they received their MMR vaccine. At a press conference, Dr. Wakefield intimated that the MMR vaccine may be a cause of autism disorders and an associated bowel disease, and he came out recommending that the MMR shot be split into individual vaccinations for measles, mumps and rubella.

The paper, and subsequent review papers by Dr. Wakefield about his findings, were seized by the media, and there was widespread panic about the safety of the vaccine, in spite of other scientific evidence that supported the safety of the MMR vaccine (many subsequent studies found no statistical evidence of MMR causing autism or bowel disease). Use of the vaccine dropped off significantly in the United Kingdom—in 1996, 92% were vaccinated, while only 84% were in 2002, with some places in London at only 61%. Similar drops in MMR vaccination rates appeared in the U.S., Europe and elsewhere around the world, because of Wakefield's work.

Figure 2.14: Dr. Andrew Wakefield

The Wakefield research was a case study in how to mislead with statistics. First and foremost, the research was not based on sound statistics. The sample under study was extremely small (n = 12) and was far from a randomized control trial—there wasn't even a control group. Much of the evidence was anecdotal, based on the recollection of a small number of parents and doctors. And there was evidence that not only was the data inaccurate but "manipulated" to imply a link between the MMR vaccine and autism. But the problems with the research were much deeper than just the statistics:

- *Dr. Wakefield had been paid the equivalent amount of over £430 000 (roughly about 800 000 in U.S. dollars) by British lawyers who were attempting to prove in court that the MMR vaccine was hazardous.*

- *There was evidence that Dr. Wakefield had applied for a patent for a competitive vaccine to the MMR that he was developing.*

- *An investigative reporter uncovered that Dr. Wakefield had planned, with the father of one of the boys in the study, to create a company that would produce testing kits for diagnosing autism, based on Wakefield's research. The "good" doctor estimated he could make over $43 million from the kits.*

In the end, after an investigation in Britain, The Lancet retracted Wakefield's research article, and in 2010, Dr. Wakefield was barred from practicing medicine ever again. But the damage had been done. You might think that there would be few ramifications to having fewer people having the MMR vaccine, but sadly that was not the case. Both measles and mumps resurged in the population. In 1998, just as Wakefield's research was published, there were 56 cases of measles, but between the start of January and the end of May in 2006 there were 449 cases, including the first death from measles since 1992. In just the month of January of 2006, there were over 5000 reported cases of mumps. And there were some serious cases, and permanent disabilities, from encephalitis (an inflammation of the brain) caused by measles. Similar patterns followed in other countries where Wakefield's research was amplified through the media. Time magazine, in 2012, listed Dr. Wakefield in their "Great Science Frauds" article.

In the end, there was a huge cost worldwide to the null hypothesis—that the MMR vaccine was safe—and the rejection wasn't even due to proper mathematics. It's yet again an indication why the bar has been (and needs to be) set so high by statistics for changing the status quo.

2.4 Review Exercises

Exercise 2.4.1. For the data $4, 6, 4, 3, 2, 5, 4, 3$ find the mean, median and mode.

Exercise 2.4.2. The following is a list of weights (in pounds) for students in a class:

$$145.6, 132.8, 138.1, 175.7, 130.2, 124.5, 115.9, 152.0, 159.3.$$

Find the mean, median and mode.

Exercise 2.4.3. The SAT tests were intended to have a mean of 500 on both the reading/verbal and math scores. Here is some data of the mean SAT reading/verbal scores in each of the years from 2003 to 2013:

year	2003	2004	2005	2006	2007	2008	2009	2010	2011	2012	2013
mean score	507	508	508	503	502	502	501	501	497	496	496

(a) What is the mean to these mean scores?

(b) What is the standard deviation?

(c) Do you think that the mean is 500 throughout the years? Explain.

Exercise 2.4.4. The following data is taken from assignment scores in a music class:

$$55, 96, 53, 76, 88, 29, 65, 72, 92, 65, 75, 68, 73.$$

(a) Find the mean, median and mode.
(b) To what percentile does the value 68 correspond?
(c) The value 75 is at what percentile?
(d) Draw a box plot of the data.
(e) Find the first and third quartiles.
(f) Find the range and the interquartile range.
(g) Find the five-number summary.
(h) Are there any outliers? Explain.

Exercise 2.4.5. Decide what kind of sampling (simple random, stratified, systematic or cluster) is described:
(a) From the population of Olympic summer athletes, the entire Nigerian team is selected.
(b) From the population of Olympic summer athletes, every 100th player entering the stadium is selected.
(c) From the population of Olympic summer athletes, from each country's team, 10 are selected randomly.

Exercise 2.4.6. Suppose we have found that in a random sample of 100 people, 64 of them have bought a new computer in the last year.
(a) What would be your point estimate for the proportion p of people who bought a new computer in the last year?
(b) Calculate a 90% confidence interval for p. What is the margin of error?
(c) Calculate a 95% confidence interval for p. What is the margin of error?
(d) Calculate a 99% confidence interval for p. What is the margin of error?

Exercise 2.4.7. The same Gallup poll as described in Exercise 2.2.13 asked the participants of the survey "And from what you have heard or read, do you believe increases in the Earth's temperature over the last century are due more to the effects of pollution from human activities (or) natural changes in the environment that are not due to human activities?" (The word "or" is in parentheses as the order between the two choices was flipped randomly to avoid bias due to position.) The poll received the following responses:

Human activities	57%
Natural causes	40%
No opinion	3%

If the margin of error is 4% at the 95% confidence level, do you think that it is likely that an equal proportion of people believe that increases in the Earth's temperature are due to human activities as believe they are due to other causes?

Exercise 2.4.8. Suppose an experiment for the effect of a new arthritis medication has been carried out. The null hypothesis is that the medication works no better than a placebo. What can you say in each of the following cases?

(a) The *p*-value turns out to be 0.0495.

(b) The *p*-value turns out to be 0.0012.

(c) The *p*-value turns out to be 0.5730.

Exercise 2.4.9. Describe the type of bias that has entered or may have entered the statistical process:

(a) For a random sample of teenagers, you advertise online for participants.

(b) An interviewer has a list of selected households to poll, but avoids ones where the porch light is out (and thus people are unlikely to be home).

(c) The mean income for staff (janitorial and administrative) at a bank is quite high.

Exercise 2.4.10. A study shows that there is a high positive correlation between the use of canes and hearing loss. Should you conclude that the use of canes causes hearing loss? Should you conclude that hearing loss requires the use of a cane?

Visualizing with Mathematics

It was the spring of 1996 and more than flowers were blooming at Stanford University. Two graduate computer science students, Larry Page and Sergey Brin, were working on a research project on the developing World Wide Web. Larry, an engineering graduate of Michigan, was introverted, while Sergey, a natural math genius who was born in Russia and emigrated to the U.S. at age 6, was more outgoing. The boys had "friended" each other, even without the help of Facebook. They had met the year earlier when second year student Brin led a bunch of potential recruits, including Page, around campus. Like all great friendships, theirs began with each of them feeling the other was obnoxious! But that didn't stand in the way of their common mathematical interests—which included the structure of the growing World Wide Web.

The Web had started out as a "universities-only" club in the late 1980s and early 1990s, but it was starting to take off among commercial organizations and companies. While less than 1% of the world was on the internet, the writing on the wall was clear—the World Wide Web was the place to be.

Page had already begun studying the important "backlinks" of web pages, the links that linked <u>into</u> a web page rather than from it, as he realized that while anyone could link to other web pages from their own web page, the backlinks that linked into the page were much more interesting and valuable. And if a really great web page linked to yours, that would be wonderful to know!

Page had already named his work "BackRub," and there was a daunting problem of trying to crawl over the web, even though it was only about 10 million pages at the time, to find the backlinks. Brin had been looking for a good project to sink his mathematical teeth into, and Larry's BackRub was the most exciting one he knew.

Once they had the web crawling algorithm working, they settled in on another fundamental problem—how to rank the pages of the World Wide Web? Certainly, the most important pages were those that were linked to from the most important pages, but that seemed circular reasoning.

However, the premise seemed sound. So how to proceed? In the end they were after a mathematical model that would provide rankings, actual numbers for each web page. It would take some thought, some analogy, some visual insight to picture what was going on in this vast network. They likely spent many a day thinking on a nearby California beach. If only the answer were as easy as a surfer hanging ten!

Mathematics can be an abstract endeavor. But we are visual beings, and sight gives a natural understanding of objects. So it is not surprising that ever since people began thinking about mathematics, they began to draw pictures—pictures to summarize, pictures to encode, pictures to analyze. The types of images that are created vary greatly, depending on what we are trying to "see" in the mathematics, and incorporate basic natural geometric objects, such as points, lines, rectangles and circles. And sometimes the key is not so much the picture that we draw, but the process involved and the intuition that we derive from connecting the mathematics to real life.

3.1   *Seeing Data*

NUMBERS ARE GREAT, BUT FOR MOST PEOPLE, YOU NEED THE NUMBERS TO COME ALIVE. In the last chapter, we saw how mathematics, in the guise of statistics, was useful in summarizing and analyzing large reams of data. The process often involved reducing one set of numbers down to a few that appeal to the mind. But can we make use of our senses, and in particular our heightened sense of sight, in comprehending the values involved? This is a purely human approach to take. Vision allows us to understand and manipulate all sorts of information in our surroundings.

Belly up to the Bar

In the previous chapter we introduced one visualization of data, **frequency plots**, which plotted vertical rectangles whose heights corresponded to the frequency for which a number occurred in a data set; these worked well, as height is a natural indicator of value. More generally, **bar charts** use rectangles (or *bars*) of equal width to illustrate values for different categories of items, with the height or length of the bar being proportional to the value for the category (bar charts can be drawn with the bars vertical or horizontal, depending on taste; if the labels for the bars are rather long, horizontal bar charts are more suitable). Usually, the bars are drawn with a small amount of space between them, though for **histograms**, the space is omitted, as the variable corresponding to the horizontal axis can take on any real value.

Frequency plots are bar charts, with the value for each category being its frequency. For example, for a list of test scores, out of 10, for a class of 20 students:

$$6, 6, 3, 8, 7, 6, 7, 5, 2, 3, 4, 5, 4, 9, 6, 7, 8, 8, 10, 5$$

the frequency table is given by:

score	0	1	2	3	4	5	6	7	8	9	10
frequency	0	0	1	2	2	3	4	3	3	1	1

A bar chart for the data is shown in Figure 3.1. (There is a choice to be made as to where to place a rectangle for a number; we choose to place it so that it meets the number on the horizontal axis at its *left* edge.)

Bar charts can be used for categories other than numbers. For instance, the *World Happiness Report 2013* tabulated the average happiness (on their scale, based on surveys, from 0 being "the worst possible life" to 10 being "the best possible life"). Table 3.1 shows the results based on regions of the world, and Figure 3.2 shows a corresponding bar chart. Which do you find more compelling and understandable—the table or the graphic?

These bar charts are **lossless representations** of the data— no information is lost from the image, and you could, if you wished or needed, reconstruct the original data from the image. Sometimes there is a lot of data for a bar chart, perhaps too much, as in the case of the test scores of Section 2.1. A bar chart

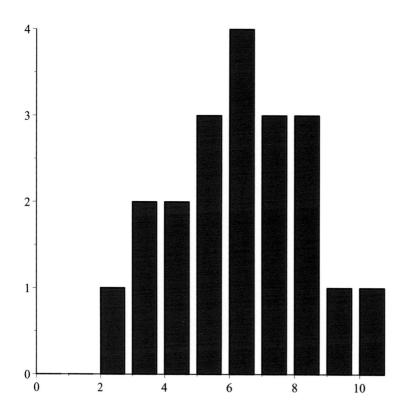

Figure 3.1: Bar chart.

Region	Happiness Score
North America, Australia and New Zealand	7.133
Western Europe	6.703
Latin America & Caribbean	6.652
Southeast Asia	5.430
Central and Eastern Europe	5.425
Commonwealth of Independent States	5.403
East Asia	5.017
Middle East & North Africa	4.841
South Asia	4.782
Sub-Saharan Africa	4.626

Table 3.1: Table of world happiness by region (from the *World Happiness Report 2013*).

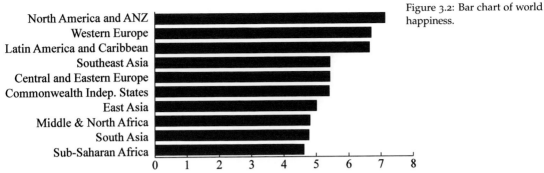

Figure 3.2: Bar chart of world happiness.

for those scores is shown in Figure 3.3. There are so many bars it is hard to gather information about the test. However, we can choose to group the data. Figure 3.5 has grouped them into groups of width 10, while Figure 3.4 has grouped them into groups of width 5 (for these bar charts, we have elected each bar to include the right edge value, as this seems most natural visually).

Such bar charts are **lossy**, as some information is lost—you can't reconstruct the original data from either of these bar charts. But what you lose in information you gain in simplicity; for both Figure 3.5 and Figure 3.4 you can see that the distributions essentially climb, then drop off, with the peak toward the end. Such a general observation gets rid of some of the "noise" in the specific values and focuses more on the trend. Of course, the bar chart in Figure 3.4 has more information than that in Figure 3.5; the choice of how to group your categories is up to the person wanting to present the data and his or her goals in doing so (more about this later).

Drawing the Line

Bar charts are an excellent visual aid when you want to compare data values for different categories. But sometimes you want to emphasize the *trend* of the data over time. In such a case, it may be better to plot a **line graph** (or **line chart**), where you plot points on a graph and connect them by straight line

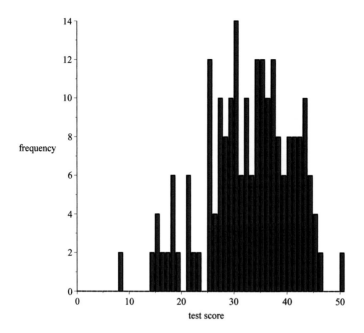

Figure 3.3: Bar chart of all test scores.

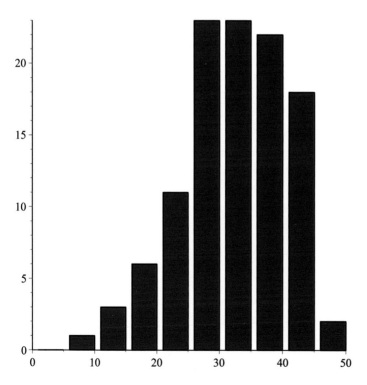

Figure 3.4: Bar chart of all test scores, in groups of width 5.

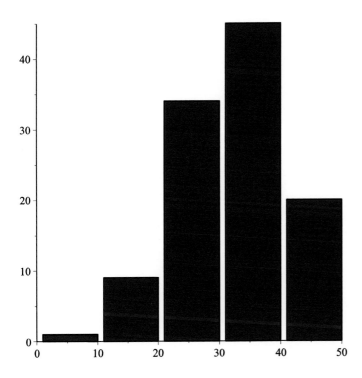

Figure 3.5: Bar chart of all test scores, in groups of width 10.

segments. For example, the data of the home runs hit by home run leaders from 1962 to 2001, given in Table 2.2, is shown with a line graph in Figure 3.6.

Line graphs are excellent for seeing trends over time. In Figure 3.6 we can see that up to 1997, the values vary a fair bit, what seems to be randomly, around 45 home runs per year, but the number of home runs hit by the leaders in 1998, 1999 and 2001 are way outside the norm. We point out that the vertical axis starts at the lowest value in the data, rather than at 0. This is a matter of choice and emphasizes here the range (and the outliers). Sometimes this is appropriate, sometimes not, as we'll see in the next section.

A Piece of the Pie

Bar charts and line graphs are very good when one wants to illustrate the changes in value between categories and over time, and where the actual values are of prime importance.

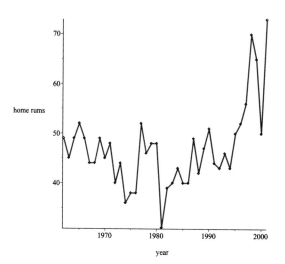

Figure 3.6: Line graph of home runs hit by home run leaders, 1962–2001.

But sometimes you want to highlight the differences of data in comparison to one another and the whole, and here another visual aid comes to the forefront. A **pie chart** represents the data as "slices" of a circular pie. The whole pie represents 100% of the data, with the area of each slice being proportional to the proportion of the group to the whole set of data. As we can naturally compare areas by eyesight, pie charts are excellent ways to make comparisons.

The area of the slices is also proportional to the central angles of the slices, as the area of a slice with central angle θ, measured in *radians*, is θr.

Pie charts work well for polling data. For example, the data in Table 2.5 gives the proportions of people in the sample who gave particular answers to the question: "Which of the following statements reflects your view of when the effects of global warming will begin to happen?"

Already begun	54%
Within a few years	3%
Within your lifetime	8%
Not within your lifetime, but affect future	16%
Will never happen	18%
No opinion	2%

A pie chart for the data is given in Figure 3.7.

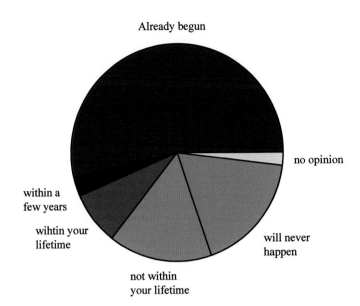

Already begun

no opinion

will never
happen

not within
your lifetime

wihtin your
lifetime

within a
few years

Figure 3.7: Pie chart of poll in Table 2.5.

When It's Good to Be Scattered

There are times when what you want to compare are two different types of numbers, in order to find a relationship among them (or a lack of a relationship between them). For example, I gathered up midterm

| 68 | 86 | 77 | 65 | 97 | 78 | 78 | 59 | 73 | 75 | 71 | 57 | 88 | 85 |
| 69 | 80 | 75 | 82 | 65 | 67 | 82 | 85 | 61 | 75 | 88 | 67 | 83 | 66 |

and final exam

51.5	76	62	51	85.5	58	64.5	57	65.5	66	36.5	44
72	58.5	72.5	59	39	46	56.5	47	66	87	44.5	46.5
58.5	48.5	61	55.5

scores for a class I have taught, in order to see if there is any relationship between them. To do so, I form what is called a **scatter plot**, which is simply a plot of points, where one variable is on the horizontal axis, the other on the vertical axis. In Figure 3.8 we see a plot of the final exam versus midterm marks. Often in a scatter plot, one restricts the axes to just include the data points, as shown in Figure 3.9.

Scatter plots are very helpful in deciding whether there is some relationship between the first variable and the second, and in particular, if there is a **linear relationship** between the two— do the points lie on a line? Lines are the one particular type of curve that everyone can intuitively visualize. For example, the points in Figure 3.9 do seem to form a line, though not perfectly. A "best line" through the points is shown in Figure 3.10. **Linear regression** is all about whether points form a line, and if so, what the best fitting line is.

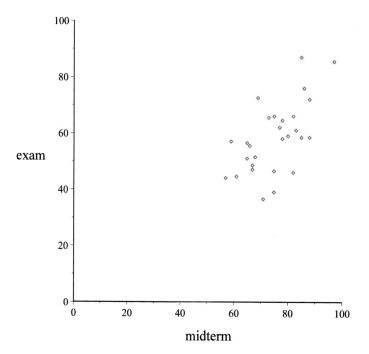

Figure 3.8: Scatterplot.

But sometimes the data, while not forming in any sense a straight line, do appear to follow other types of curves. For example, in the study of **aesthetics** (beauty), data like that shown in Figure 3.11 has been observed when asking individuals to rate how complex a piece of music is compared to how much they like it (the data shows average ratings on pre-designed scales for complexity and likability, for a number of different songs).

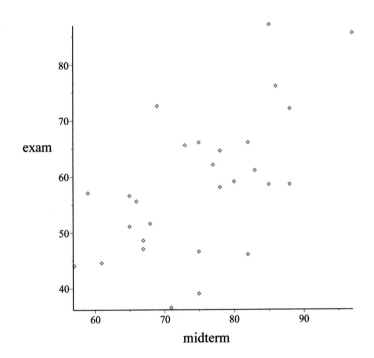

Figure 3.9: Scatterplot, restricted to include just the data points.

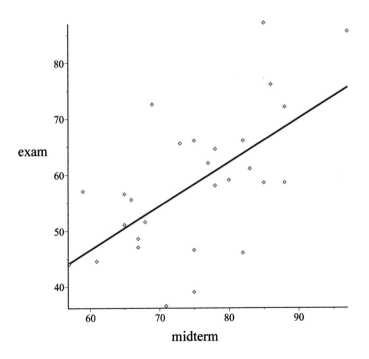

Figure 3.10: Scatterplot with a best fitting line.

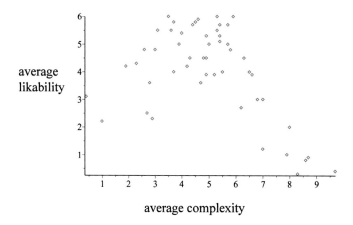

Figure 3.11: Scatterplot of likability versus complexity for different musical pieces.

Without a scatter plot, we would be hard pressed to say anything about the data. From the scatter plot we immediately see that there is no linear relationship between likability and complexity, but we do see the outline of a curve, as shown in Figure 3.13. This curve is called, for obvious visual reasons, an **inverted U curve**, and points to the fact that people tend to like music (and indeed other forms of art like paintings and poems) that is not too simple and not too complex—something in the middle is found to be most desirable, though where the middle is varies from person to person. The inverted U curve is also called the **Wundt curve** and appears in numerous places throughout the arts.

Figure 3.12: Psychologist Wilhelm Wundt (1832–1920).

Life Lesson: When it comes to aesthetics, people most like art that is not too simple and not too complex.

And the List Goes On . . .

There are many other choices for presenting data visually. Some are based on the ones already described. For instance, you can replace items like points, lines, rectangles with pictures to form **pictographs**. For example, a pictograph corresponding to

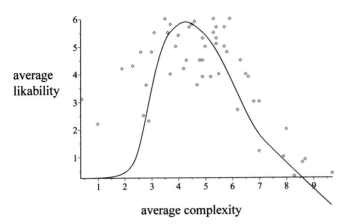

Figure 3.13: Wundt (inverted U) curve with data points.

the happiness survey in Table 3.1 is shown in Figure 3.14. Things can be further spiced up, if you wish, with three-dimensional graphs as well.

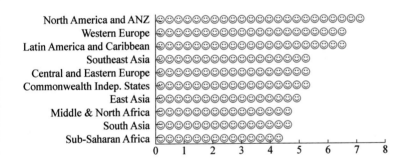

Figure 3.14: A pictograph.

Some of the other types of visualizations of data deal with three variables at a time, rather than two. When these variables are all real numbers, one can plot in three-dimensional space, but the problem with that is that many people have little intuition in three dimensions. The are many other ways to represent a third dimension without having to go into three-dimensional plots—one can make the third dimension color, intensity, and so on. One that I like is called a **bubble plot**, where you plot the first two variables in the usual way, on the plane, and then you draw a bubble (that is a circle) whose radius is proportional to the third variable's value (if the value is positive, you draw a filled circle, if it is negative, you draw an open circle). So, for example, if we were tabulating the occurrence of a disease in a city, with its location from the city center measured east-west and north-south in miles (west and south are given as negative values, east and north are positive), then we might collect data as the following:

East-West (miles)	North-South (miles)	Number of Cases
3.2	0.5	4
3.7	1.1	5
-2.5	8.2	2
-1.8	-2.6	4
5.8	6.7	6
3.2	-4.3	2
5.3	4.8	3
-1.4	5.2	3
-0.2	-1.5	5
1.4	7.3	2

A bubble plot of the data is shown in Figure 3.15. Note that from the plot we see that the disease likely started in the upper right quadrant, mostly tracking down south-westerly through the city, leaving pockets of infected cases along the way. Would that be noticeable without the plot? Perhaps, but most likely not. The eyes really do have it!

Visual Trickery

Just as one can be taken down the garden path with statistics, one can also be misled, perhaps even more easily, with graphics, as we feel that our eyes never lie! Here are but a few ways to be misled by mathematical pictures:

1. **Presenting the wrong plot.** The type of plot chosen for given data is dependent on what type of point the presenter wishes to make. So, for example, it would be misleading to present the following pie chart for the world happiness data in Table 3.1, even though it is a perfectly good pie chart. The issue

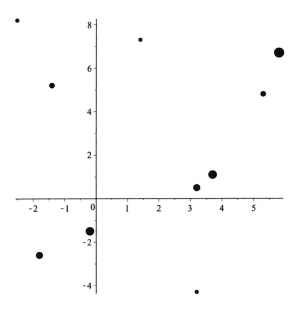

Figure 3.15: A bubble plot.

is that by using a pie chart, the implication is that we are asking the reader to implicitly consider each region's happiness in comparison to the "whole" happiness of the world, where it makes little sense to define the latter as the sum of the happiness of the regions! Also, the pie chart hides the values of happiness, so that the reader can't decide whether each region is basically happy or unhappy.

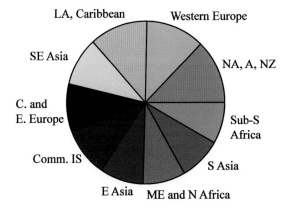

2. **Leaving out or inserting axes.** In some types of plots, one has a choice whether to include or exclude the origin, and the decision can affect the viewer's perception of the data. For example, here are two line graphs that both show the same data, namely, income levels over five years. The first picture seems to show great differences (and an increasing trend) in the income levels, while the second one, whose vertical axis goes down to 0, shows little change in the data, which is probably closer to the truth about what is going on.

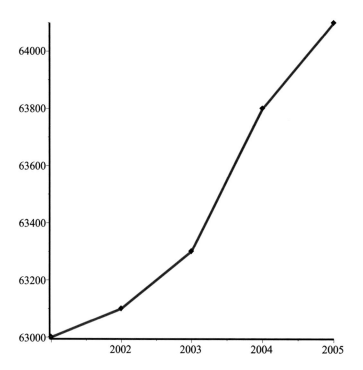

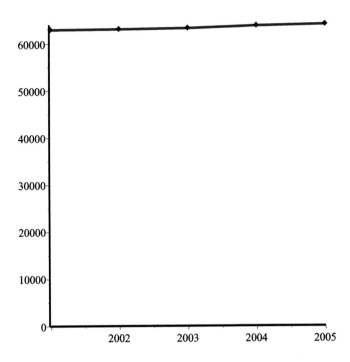

3. **Covering up missing data.** If data is missing, any relevant graphic should show that fact. For example, suppose that heart rates for an individual in hospital were recorded over a twelve hour shift, from 07:00 to 19:00 (that is, 7 a.m. to 7 p.m.):

Time	Heart Rate
07:00	62
08:00	53
09:00	66
10:00	61
14:00	123
15:00	118
16:00	136
17:00	142
18:00	133
19:00	129

The first line chart would be incorrect, as it hides the fact that the heart rates were not recorded at 11:00, 12:00 or 13:00, and this may be significant. Did the heart rate climb gradually over time? How long has the heart rate been high? This is called *tachycardia*, and that may be very important to know, or to know that you don't know! The second chart would be more informative (and truer to the data, or lack thereof).

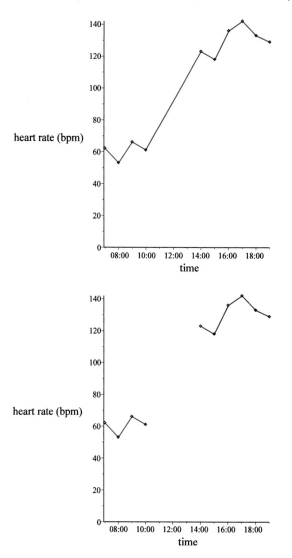

4. **Cute but misleading pictographs.** Pictographs are often eye catching, but they can confuse, unintentionally or intentionally. For example, we naturally correlate visually taller objects

as more. So in the following two side-by-side pictographs, the six bundles on the left should be seen as more than the five stacked on the right, but the one on the right seems to indicate a larger number than that on the left as it is taller.

We also have a natural inclination to see a larger area as indicating more. In the following pictograph, we see the stack on the right as being "more" than the stack on the left, simply by the fact that it takes up a larger area.

You may think that the examples mentioned are cooked up and don't actually occur in practice. But here is an example of a graphic similar to one used on a major TV network.[1] Clearly the heights of the bars should be proportional to the values (6 000 000 and 7 066 000) but aren't, and our natural, unavoidable inclination to see heights as being proportional to the values misleads us to believe that the number of Americans enrolled in Obamacare as of March 27 is *less than half* of the goal as of the end of March, when this is far from the truth.

[1] This graphic is based on a figure originally shown on Fox News.

OBAMACARE ENROLLMENT

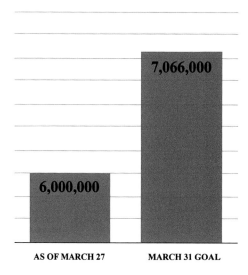

A misleading graphic!

Why was the graphic so misleading? It's hard to know, though the television station is known for its right-wing slant on the news.

OBAMACARE ENROLLMENT

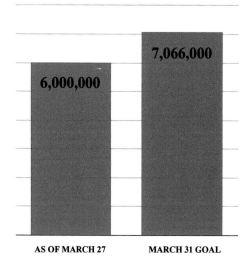

A corrected graphic for Obamacare.

Exercises

Draw a bar chart for the data in each of the following four exercises. For each. What can you say by looking at the bar chart?

Exercise 3.1.1. 7, 2, 7, 5, 3, 2, 8, 5, 9, 7, 6, 7, 7, 4, 10, 1, 1, 8.
Exercise 3.1.2. 9, 2, 9, 7, 7, 3, 8, 7, 9, 7, 6, 7, 7, 9, 10, 9, 9, 8.
Exercise 3.1.3. 1, 2, 9, 5, 6, 3, 4, 2, 2, 2, 1, 4, 3, 5, 7, 5, 1, 3.
Exercise 3.1.4. 9, 2, 9, 7, 1, 3, 2, 7, 9, 7, 8, 7, 7, 9, 2, 2, 3, 8.

Exercise 3.1.5. Here are some test scores (out of 100):

51.5	76	62	51	85.5	58	64.5	57	65.5	66
36.5	44	72	58.5	72.5	59	39	46	56.5	47
66	87	44.5	46.5	58.5	48.5	61	43		

(a) Draw a bar chart for the data with rectangles of width 5. What do you see about the data?
(b) Draw a bar chart for the data with rectangles of width 10. What do you see about the data?

The following table (taken from the U.S. Department of Transportation's *Traffic Safety Facts: 2010 Data*) shows the number of motor vehicle deaths per 100 000 people in the U.S., and per 100 000 registered vehicles, from 2001 to 2010:

Year	Fatality rate per 100 000 pop.	Fatality rate per 100 000 reg. vehicles
2001	14.81	19.07
2002	14.95	19.06
2003	14.78	18.59
2004	14.63	18.00
2005	14.72	17.71
2006	14.31	16.99
2007	13.70	16.02
2008	12.31	14.43
2009	11.05	13.08
2010	10.63	12.64

Use these tables to answer the following six problems.

Exercise 3.1.6. Draw a line graph of the fatalities per 100 000 population over the time period, starting the vertical axis at 0.

Exercise 3.1.7. Draw a line graph of the fatalities per 100 000 population over the time period, with a vertical range from 10.00 to 15.00.

Exercise 3.1.8. Explain what you could emphasize with each of the graphics in Exercises 3.1.6 and 3.1.7.

Exercise 3.1.9. Draw a line graph of the fatalities per 100 000 registered vehicles over the time period, starting the vertical axis at 0.

Exercise 3.1.10. Draw a line graph of the fatalities per 100 000 registered vehicles over the time period, with a vertical range from 10.00 to 20.00.

Exercise 3.1.11. Explain what you could emphasize with each of the graphics in Exercises 3.1.9 and 3.1.10.

The following tables (also taken from the U.S. Department of Transportation's *Traffic Safety Facts: 2010 Data*) show the number accidents where speeding was a factor, from 2001 to 2010:

Year	Number of speeding accidents	Percent of accidents involving speeding
2001	12 924	31
2002	13 799	32
2003	13 499	31
2004	13 291	31
2005	13 583	31
2006	13 609	32
2007	13 140	32
2008	11 767	31
2009	10 664	31
2010	10 395	32

Use this table to answer the following eight problems.

Exercise 3.1.12. Draw a line graph of the number of speeding accidents over the time period, starting the vertical axis at 0.

Exercise 3.1.13. Draw a line graph of the number of speeding accidents over the time period, with a vertical range from 10 000 to 14 000.

Exercise 3.1.14. Explain what you could emphasize with each of the graphics in Exercises 3.1.12 and 3.1.13.

Exercise 3.1.15. Draw a line graph of the percent of accidents involving speeding over the time period, starting the vertical axis at 0.

Exercise 3.1.16. Draw a line graph of the percent of accidents involving speeding over the time period, with a vertical range from 31 to 32.

Exercise 3.1.17. Explain what you could emphasize with each of the graphics in Exercises 3.1.15 and 3.1.16.

Exercise 3.1.18. What explanation can you give for why the number of speeding accidents *decreased* from 2009 to 2010 and yet the percentage of accidents involving speeding *increased*? Explain your answer.

Exercise 3.1.19. About how many accidents were there in 2001? In 2010?

A Gallup poll taken March 6–9, 2014 of 1048 adults, asked the following question: *Thinking about what is said in the news, in your view is the seriousness of global warming generally exaggerated, generally correct, or is it generally underestimated?* (Answers were rotated among respondents to ensure the order did not affect the outcome.) (Copyright ©2014, Gallup Inc. All rights reserved. The content is used with permission; however, Gallup retains all rights of republication.)

	Republicans %	Democrats %	Independents %
Generally exaggerated	68	45	18
Generally correct	15	21	32
Generally underestimated	15	32	49

Use this table to answer the following five problems.

Exercise 3.1.20. Why don't the numbers in each column add up to 100?

Exercise 3.1.21. Draw a pie chart for the Republican response.

Exercise 3.1.22. Draw a pie chart for the Democratic response.

Exercise 3.1.23. Draw a pie chart for the Independent response.

Exercise 3.1.24. What conclusions can you draw from the three pie charts?

A group of students gathered data on income and happiness from 15 subjects (income was measured in yearly dollars, while happiness was recorded on a scale from 0 to 100):

Subject number	Income (yearly)	Happiness
1	86 000	75
2	34 000	52
3	52 000	58
4	55 000	64
5	29 000	50
6	14 000	45
7	102 000	78
8	77 000	75
9	147 000	97
10	24 000	54
11	38 000	60
12	95 000	82
13	33 000	38
14	48 000	65
15	72 000	71

Use this table to answer the following two problems.

Exercise 3.1.25. Draw a scatter plot for the data, with happiness (vertical axis) versus income level (horizontal axis).

Exercise 3.1.26. Do you think the data visually forms a straight line? if so, find the equation of a well-fitting line to the data.

Data was collected on the number of reported cases of a certain bacterial infection in an urban location. The following table lists the location (from the city center) along with the number of cases:

East-West (miles)	North-South (miles)	Number of Cases
3.2	0.5	9
3.7	1.1	5
-2.5	8.2	8
-1.8	-2.6	3
5.8	6.7	7
3.2	-4.3	4
5.3	4.8	5
-1.4	5.2	3
-0.2	-1.5	2
1.4	7.3	10

Use this table to answer the following two problems.

Exercise 3.1.27. Create a bubble plot for the data.

Exercise 3.1.28. From the plot, what conclusions can you draw about the course of the infection?

Exercise 3.1.29. A pie chart for the percent of accidents involving speeding from 2001 to 2010 (see the table preceding Exercise 3.1.12) is shown below. Is it appropriate? Explain your answer.

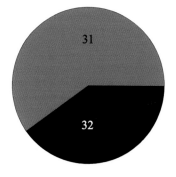

Exercise 3.1.30. The following line graph shows a baseball player's batting average over a 5-year period. Why is the graphic misleading?

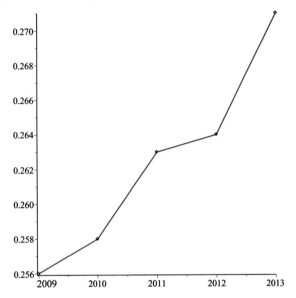

Exercise 3.1.31. What is wrong with the following graphic?

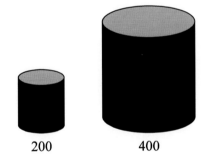

200   400

2   *A Graph Is Worth a Thousand Words*

Visualization is important for much more than just numbers—it is very helpful to "see" relationships between objects. And this brings us to the fascinating and oh so useful topic of **graphs** and **networks**. These graphs are unlike the ones you would see when you plotted functions, though they are named the same and both are visualizations. Let's start off with the definition.

124

Definition 3.2.1. A **graph** or **network** is a collection of objects, called **vertices**, together with a relation between objects—for two objects x and y, either x is related to y (in which case we say that there is an **edge** from x to y), or they aren't.

While the definition doesn't mention any pictures, all mathematicians who work with graphs draw pictures, and it is indeed the key to graphs' aesthetic appeal. When one draws a graph, the vertices are drawn as points on the page, and the edges are shown with line segments or curves (if the relation is symmetric, that is, if x relates to y just in case y relates to x), and with arrows additionally otherwise. The shape of the curve makes no difference—all that matters is which vertices are joined to which other ones. Figure 3.16 shows two examples of graphs. The one on the left shows an **undirected graph**, where a line between two vertices indicates that they are **adjacent**, that is, related to each other in both directions; the edge (that is, line segment) between vertex a and vertex c indicates that a is related to c and c is related to a. The one on the right is a **directed graph**, where there are some vertices that relate to one another in only one direction.

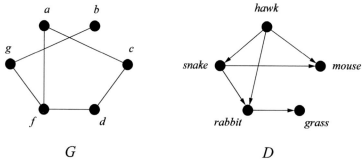

Figure 3.16: Examples of graphs.

G   D

The graph on the right is directed—the edges have arrows on them. This directed graph represents part of a **food web**, with the arrows from animal x to animal y if animal y is a predator of animal x (that is, x is a prey of y). So, for example, hawks are predators of snakes, mice and rabbits, but not vice versa. Of course, we could describe a graph just mathematically or just in words, but that would remove the intuition we have with the pictures.

As with food webs, graphs arise in so many applications.

The key in each is how the visualization helps you answer an important problem. Here are just a few:

- You can take the street intersections as your vertices and the roads as your edges. This can be a directed graph if some of the roads are one-way streets. One important problem is how to navigate around in the city, from one destination to another, quickly (such graphs are utilized in routing in conjunction with GPS systems). The first picture is a map of the downtown core. The second graph is a mix of directed and undirected edges, corresponding to one- and two-way streets. The vertices are the intersections, which I have labeled with i subscripted by the avenue and street numbers. Notice how the graph gets rid of inessentials, such as the buildings on the streets, the lengths of the roads, and so on—all that is important for our problem is the layout. This is mathematical graphics at the best—highlighting what is important to us, hiding the rest, in order for us to "see" the solution!

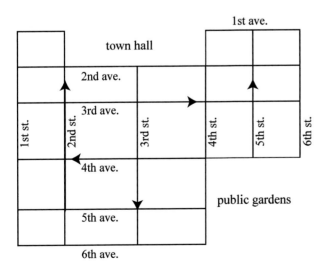

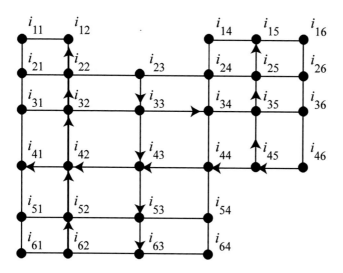

- Undoubtedly, when you were young, you were given a picture like the one below at the left and asked to trace the figure, without lifting your pen from the page and without retracing any line. If we draw a vertex at every intersection of two line segments, we get a graph, and the problem converts into one where we ask whether we can, starting at a vertex, travel throughout the graph, tracing over every edge exactly once. That may not seem like a big step, but we'll see soon how big a leap in understanding that is.

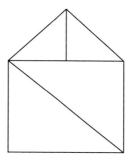
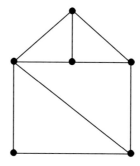

- For a large scale graph, you can take all adults in the U.S. as your vertices, with edges denoting pairs who know each other—this yields an undirected graph, an **acquaintance graph**, which is an example of a **social network**. An interesting question is how far apart two people are—if you wanted to reach out to a random person in America, how long a chain of acquaintances would you have to use? This graph is much too big to draw (it has over 220 million vertices!), but we can schematically think of the graph by drawing an image that *suggests* the entire graph.

There are a lot of properties of social networks that can be explored using graphs. For instance, one might be interested in who the most important people are. One measure could be who has the most acquaintances, which corresponds to a vertex of biggest degree in the graph (the **degree** of a vertex in an undirected graph is simply the number of edges that emanate from the vertex, so that in the acquaintance graph G shown, Sarah has degree 6).

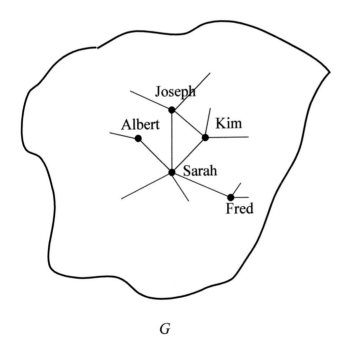

G

You could also explore groups such that every pair of people are acquaintances—in graph theory terminology, these are called **cliques**, and pictorially, they appear as a set of vertices all joined to one another. In social network G, Sarah, Kim and Joseph form a clique of size 3 (that is, a "triangle" in the graph). It's been noted that in many social networks, your friends tend to have known one another, so that there are large cliques among your friends.

The question of how long a chain is needed is a famous one, and it is believed that at most 6 steps (that is, edges) are need to reach from any one person to any other in the social network of the U.S. (or indeed the world)—this is known as the **six degrees of separation** principle.

Life Lesson: When meeting someone for the first time, talk to them—ask them about where they grew up, where they worked, where they went to school, where they've traveled. If you try, you can almost certainly find a short chain of acquaintances to connect you to them, and that will make them feel closer to you!

The theory and applications of graphs is an enormous topic. We'll only touch here on a few problems, chosen both for their importance and for interest. Definitions will crop up throughout, but they arise from the pictorial view of graphs and our experiences in real life, so they should seem natural to you.

Eulerian Graphs and Tracing Figures

Your first exposure to graphs, unbeknownst to you, was likely the question of tracing a figure without taking your pencil of the page, and without retracing any line. If you were like me, you probably spent several hours trying one or the other of these.

Look at the figure, and let's go through some preliminary reasoning. If you can trace the figure as described, then you can do so starting at one of the corner points, rather than in the middle of a line, because if you can start in the middle of one of the lines, you of course have to end back up at the end covering the rest of the line. By starting your "tour" at the first corner you

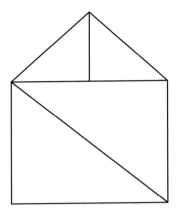

Figure 3.17: Is this graph traceable?

hit along the way, you can end back up at the corner following the same path. So the tracing, if it can be done at all, can always start at a corner. But we can make the figure into a graph by placing a vertex at each corner, so we have just translated the problem into finding a **trail** in the graph, that starts at a vertex, and travels along the edges, from vertex to vertex, without revisiting an edge (though you can, of course, pass through a vertex more than once). Such a trail is called an **Eulerian trail**, named after one of the most famous mathematicians of all time, Leonhard Euler.

A **walk** in a graph from vertex u to vertex v is a an alternating sequence of edges, starting at u and ending at v, such that each edge in the walk has a vertex in common with the next one. The walk is a **trail** if no edge is repeated and a **path** if no vertex is repeated.

The problem was brought to Euler's attention by the towns-folk of Königsberg. The city had two islands that were joined to each other and to the rest of the city by seven bridges (see Figure 3.19). Couples would attempt to traverse each bridge exactly once on their outing, returning to where they started, but many, many years went by to no avail. Euler brilliantly realized that one could replace each land mass by a vertex, and each bridge by an edge joining the relevant vertices to form a graph—see Figure 3.20 (this is cited by many as one of the first instances of a graph). Then the problem of traversing the bridges of Königsberg was translated into the search for what we now call an **Eulerian circuit** in the graph, which is an Eulerian trail that starts and ends at the same vertex.

Figure 3.18: Mathematician Leonard Euler (1707–1783).

The critical observation is the following. If a graph has an Eulerian circuit, then for every vertex v except the start and end vertex x, whenever you enter v on the tour (of course with a new edge), you leave by a new edge, so the edges out of v are paired up by the tour. This means that vertex v has even degree. By similar reasoning, the edges out of x pair up as well, with the first edge of the tour pairing up with the last edge of the tour. This means that **if a graph has an Eulerian circuit, then every vertex has even degree**. Moreover, assuming there are no **isolated vertices** (that is, no vertices with degree 0) the graph must be **connected**, that is, you can walk from every vertex to any other. So if a graph has an Eulerian circuit, it must be connected and every vertex must have even degree. This observation immediately told Euler that the Königsberg Bridges problem was not only difficult but *unsolvable*—there were two vertices with odd degrees (of 3 and 5. Likewise, the graph in Figure 3.17 cannot be traced, as it has four vertices of odd degree.

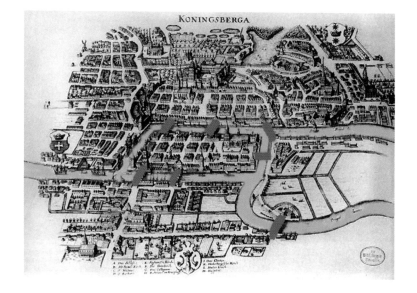

Figure 3.19: The town of Königsberg. The seven bridges are as indicated.

Now the result we have is useful, but only half of the puzzle. What about if a graph is connected and has all vertices of even degree—does it then have an Eulerian circuit? The answer is, perhaps surprisingly, yes! We won't delve into the argument (that will be hinted at in the exercises), but we will talk about a procedure, an algorithm, for finding such a circuit when it exists. It is called **Fleury's algorithm** and can easily be stated:

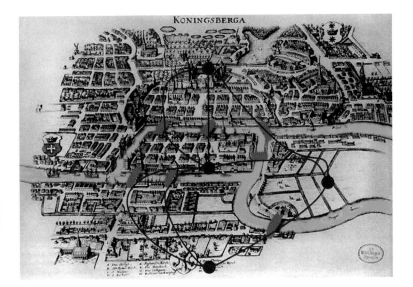

Figure 3.20: Königsberg graph.

> **Fleury's algorithm for finding an Eulerian Circuit:** Start at any vertex v and walk along a new edge as you proceed, erasing it from the graph as you walk over it (as it cannot be used again). If there is a choice of which edge to use next, always choose one that won't break up the remaining graph into more pieces, if possible (otherwise, choose any edge).

So let's give Fleury's algorithm a whirl, on the graph in Figure 3.21. As the graph is clearly connected and every vertex has even degree, we can start anywhere. Let's start at vertex a. Then we can choose to walk across to b, deleting the edge as we go along, and then go down to d, removing the edge form b to d. At this point, when we walk out of d with the other edge, we break up the graph into more pieces, as d suddenly becomes

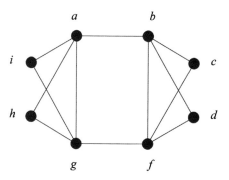

Figure 3.21: Eulerian graph.

disconnected from the rest of the graph, but there is no choice, so we do so, and erase the edge. We are now at vertex f, and we either choose to go across to g, or back to b or c. But going across to g will break the graph up into more pieces, as the removal of the edge from f to g separates b, c and f from the vertices on the left, while going to c and removing the edge from f to c does not break up the graph into more pieces (going back to b would do the same). So we do the latter, as shown in Figure 3.22 (we number the edges in the order we traverse them, and rather than erase edges as we go, we dash them so they are still visible). We continue on with the algorithm to find the Eulerian circuit shown in Figure 3.23.

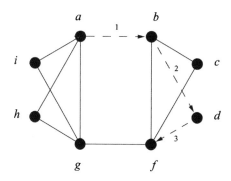

Figure 3.22: Eulerian graph, part way through Fleury's algorithm.

What about whether we can trace a figure without having to start and end at the same place? That is, whether a graph has

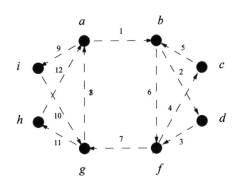

Figure 3.23: Eulerian graph, at the end of Fleury's algorithm.

an Eulerian trail? It's not too difficult to see that a graph has an Eulerian trail if and only if the graph is connected and there are at most two vertices of odd degree. If the graph is connected, then (i) if there are no vertices of odd degree, the graph has an Eulerian circuit (which is an Eulerian trail), and you can start wherever you like, and (ii) if there are exactly two vertices of odd degree, the graph has an Eulerian trail (but not an Eulerian circuit), and you can start at one of the odd vertices and end up at the other.

Now tracing figures is a trip down memory lane, but there are real-life applications of Eulerian graphs as well. Assigning a route to a garbage truck in an area of the city is an Eulerian problem one would want a route that traveled down each street exactly once (to minimize cost) and return to the depot. And the existence of Eulerian trails and circuits even has applications to DNA sequencing as well. You can be sure, one thing that mathematicians love to do is use and reuse previously solved problems!

We remark that while finding a circuit that visits every *edge* exactly once is not too hard, the problem of visiting each *vertex* exactly once, called the **Hamiltonian cycle problem**, is much, much harder, with no good way to solve it in general.

Spanning Trees and Low Cost Connections

Connectivity plays an important role in the existence of Eulerian circuits and Hamiltonian cycles, as well as in many other problems on graphs. The fewest number of edges needed to connect up a set of vertices is always one less than the number of vertices, and such a minimal connection of edges is called a **spanning tree**, as it visually resembles a tree. (In Figure 3.31, redrawn in Figure 3.24 without the vertex labels, there are 8 vertices and 7 edges in the spanning tree shown in red.) Every connected graph has a spanning tree—this follows from the fact that if an edge is in a cycle, it can be deleted, leaving the graph still connected.

Figure 3.24: Spanning tree.

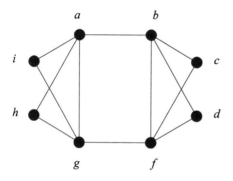

A **weighted graph** is a graph where every edge has a non-negative number attached to it. If we have a spanning tree in a weighted graph, we can talk about its *cost*, which is just the sum of the weights of its edges. Figure 3.25 shows a weighted graph, along with all eight of its spanning trees. Their weights can be calculated as follows:

- Spanning tree T_1 has weight $7 + 4 + 8 = 19$
- Spanning tree T_2 has weight $7 + 3 + 4 = 14$
- Spanning tree T_3 has weight $7 + 5 + 4 = 16$
- Spanning tree T_4 has weight $7 + 3 + 8 = 18$
- Spanning tree T_5 has weight $5 + 8 + 4 = 17$
- Spanning tree T_6 has weight $5 + 3 + 8 = 16$
- Spanning tree T_7 has weight $5 + 7 + 8 = 20$
- Spanning tree T_8 has weight $5 + 3 + 4 = 12$

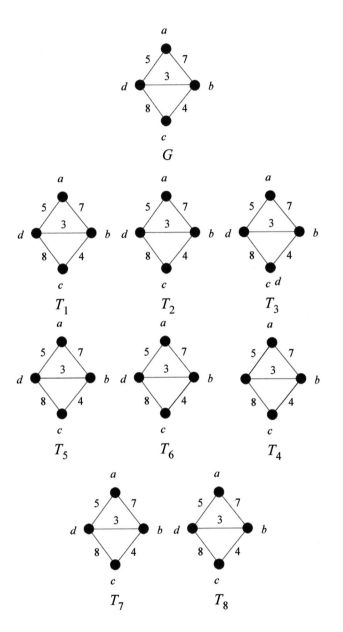

Figure 3.25: Weighted graph with its weighted spanning trees.

The minimum of all of these is 12, which occurs for the spanning tree T_8, but there was a fair bit of work involved, listing all of the spanning trees. We will be interested in finding a **minimum cost spanning tree** in a weighted undirected graph with a procedure that doesn't involve listing all of the spanning trees. Fortunately, there is such a procedure.

Prim's tree-growing algorithm for finding a minimum cost spanning tree in an undirected connected graph: Pick any starting vertex v and mark it "discovered." As long as you don't have a spanning tree, do the following:

> Among all the undiscovered vertices adjacent to a discovered vertex, pick one, say u, with an edge e of smallest weight to a discovered vertex. Mark u as discovered and color the edge e.

At the end, the colored edges will form a minimum cost spanning tree.

So let's do an example. Suppose we have a layout of possible connections we can make between offices and we want to wire the offices as cheaply as possible for the hooking up of the computers in the offices. How do we do so? Figure 3.26 show a set of offices, each with a computer, that need to be wired together.

The cost of the wiring between offices varies by the distance, the type of wire needed, the difficulty in wiring between the rooms, and so on. We'll simplify things by replacing each office by a vertex and join two by an edge whose weight is the cost of wiring the two computers directly (see Figure 3.27).

From Prim's algorithm, we're free to start where we like, so let's start in the upper right-hand office. We'll mark this vertex by coloring it red. We choose the edge with the least weight from the red (discovered) vertices to the black ones (undiscovered); there is only one choice, the horizontal edge weighted 150 out of the vertex in the top right. We color this edge, adding it to

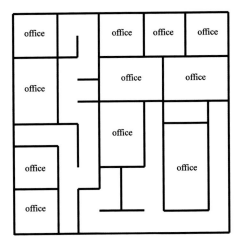

Figure 3.26: Wiring problem.

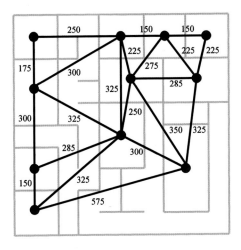

Figure 3.27: Wiring problems graph.

our spanning tree, and discovering its endpoint.

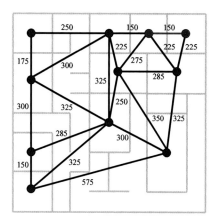

We choose the next edge to be added from the three edges that go from a discovered (colored) vertex to an undiscovered (uncolored) one, and again the horizontal edge out to the left of the last discovered vertex is the least. So we add it to our growing spanning tree and discover its other endpoint.

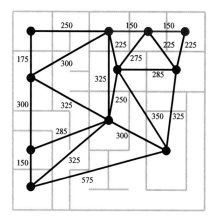

At this point, we consider all the least weight edges from a discovered vertex to an undiscovered one, and there are three such edges, each of weight 225. We arbitrarily pick one, and the edge to the spanning tree and discover its other end.

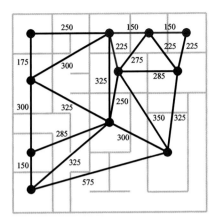

At the next step we have another choice between two edges of weight 225, so we pick one again.

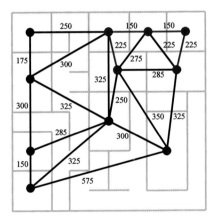

Now there is another edge with weight 225, but it is not available as it joins two discovered vertices, and not a discovered and an undiscovered pair. There are two available edges with the smallest weight, 250, so we pick one and add it.

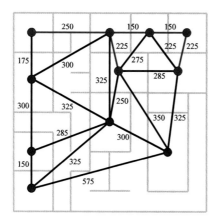

We continue in this manner until all of the vertices are discovered.

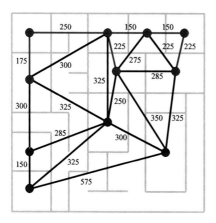

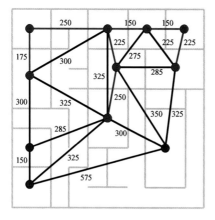

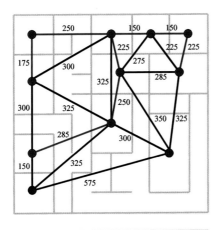

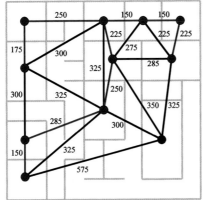

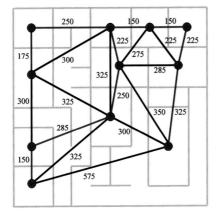

With Prim's algorithm complete, we have a minimum cost spanning tree, which is a minimum cost wiring of the offices' computers into a network.

Minimum cost spanning trees arise in many other applications. One is clustering of data points into groups—one looks at the distances between data points and grows clusters in a tree-like manner. Another is to **phylogenetic trees**; once you have a measure of distances between life forms (whether they are animals or DNA sequences) you can search for a minimum cost spanning tree that can be useful in proposing the underlying evolutionary history. Finally, minimum cost spanning trees have even found use in explaining the spread of viral diseases.

Shortest Paths and Mazes

Some puzzles you would have seen when you were younger would include **mazes**. I'm sure you spent hours trying to find your way out of a maze on a page. Here again graph theory comes to the rescue. To explain how to do it, we solve a more general problem.

We turn again to weighted graphs. Figure 3.28 shows a downtown core, with the weights on each edge being the length of the street (between the relevant intersections). If we have a graph of airplane routes, then we could label each edge with the cost of a one-way ticket for that flight (see Figure 3.29).

The questions we want to answer are the following: What is the shortest way to get from one corner to another in the downtown core? What is the cheapest way to fly from one city to another? These problems sound like they might be related to minimum cost spanning trees, but they all turn out to be part and parcel of one general different problem on weighted graphs: what is the shortest path from one vertex to another, where the length of a path is the sum of the weights on all of the edges in a path?

Dijkstra's algorithm quickly finds shortest paths from any one vertex v to all of the rest. The basic idea is this: You start at v and successively locate vertices on shortest paths from v, designating them as "discovered" along the way. The whole process revolves around the observation that if you have a shortest path from v to u, then the path from v to each intermediate vertex must be a shortest path as well.

Dijkstra's algorithm for finding shortest paths from one vertex v to all of the rest, in either a directed or undirected graph:
Start by labeling v with 0, mark it "discovered," and label all of the other vertices with ∞, the infinity symbol (we'll update this shortly). Then repeat the following until no more vertices can be marked as "discovered":

1. For each "undiscovered" vertex u that is joined by an edge from the last discovered vertex w: Update the label of u with the *smaller* of

    *   its present label or
    *   the sum of the labels at w plus the weight on the edge from w to u.

2. Mark an undiscovered vertex u with the smallest label as our next discovered vertex, and color the edge between this new discovered vertex and the found vertex y that it is joined to such that the label of y plus the weight of the edge from y to u is equal to the label at u.

At the end, the label at each vertex x is the length of a shortest path from v to x. (If x is labeled with ∞ at the end, there is no path from v to x.)

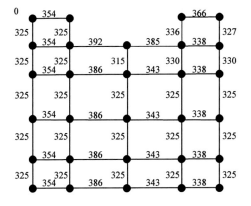

Figure 3.28: Weighted graph of a downtown core.

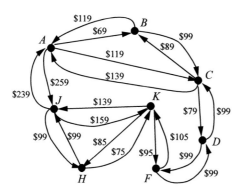

Figure 3.29: A weighted
graph of the cost of one-way
airline tickets.

We won't explain in detail why the algorithm works, though
it's not too hard (it depends mostly on the observation that a
shortest path P from v to x consists of shortest paths from v to
all vertices on P). Let's illustrate Dijkstra's algorithm with the
airline flights in Figure 3.29. Let's find the shortest path (that is,
the cheapest flight) from airport A to all others.

We start by labeling A with label 0 and label the other vertices
with ∞. A is marked as discovered (we indicate a discovered
vertex by coloring it red).

Figure 3.30: Edsger Dijkstra,
1930–2002 (courtesy: The
History of Computing Project
2014).

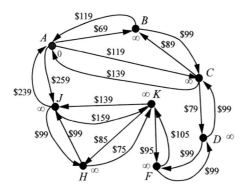

We next update the labels on all vertices adjacent to A, namely,
B, C and J, by adding the label at A, which is 0, to the weights of
the edges, namely, 119, 119 and 259, respectively, so that the label
at B becomes the smaller of ∞ and $0 + 119 = 119$, which is 119,
of course. Likewise, C and J get labels 119 and 259, respectively.
We pick an undiscovered vertex with the smallest label, which is
B or C—we arbitrarily choose B—and mark it found, and color

the edge from the discovered vertex *A* to the just found vertex *B*, as the label at *B*, 119, is the sum of the label at *A*, 0, with the weight of the edge between them, which is 119.

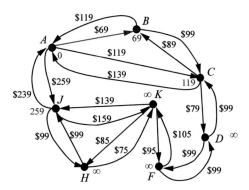

We proceed. There is no undiscovered vertex joined to *B*, so no labels change. The undiscovered vertex *C* is marked as discovered, and the edge from *A* to *C* is colored.

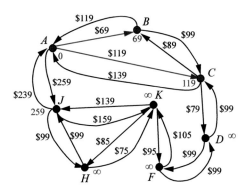

D is now the only vertex whose label is updated, and it turns out that *D* has the smallest label among all of the undiscovered vertices, so it is marked as discovered, and we color the edge from *C* to *D*.

146

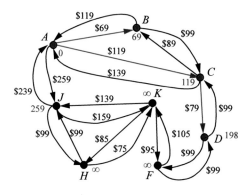

F's label is updated to 297, J now is the undiscovered vertex with the smallest label, so J now gets marked, and the edge from A to J is colored, as J's label is the sum of the label at A and the weight of the edge from A to J.

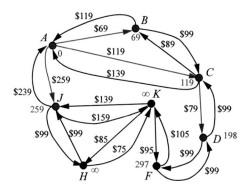

We now update the labels at H and K because of J. Then F, H and K are the only undiscovered vertices, and F has the smallest label, so it gets marked as discovered, with the edge from D to F colored.

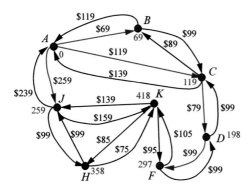

The only undiscovered vertex adjacent to the last marked vertex, F, is K, and its label, 418, is larger than the sum of the label, 297, at F plus the weight of the edge from F to K, 105. Thus we change the label at K to $297 + 105 = 402$, and H has the smallest label, so it gets marked as discovered, with the edge from J to H colored.

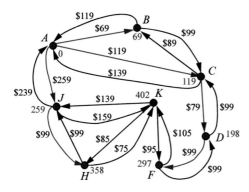

The only undiscovered vertex left is K, and its label doesn't change as $402 < 358 + 75$. Again, no change is made to the labels, and K, by default, has the smallest label, so it gets marked as discovered, with the edge from F to K colored.

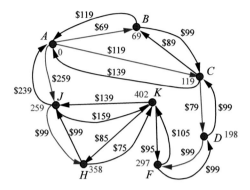

As all vertices have been marked as discovered, we are done—
the labels indicate the length of the shortest path in the graph
from A to all other vertices. So, for example, the length of the
shortest path from A to K is 402, that is, the cheapest airfare
from airport A to airport K is \$402. What about the actual short-
est path, i.e. the shortest route, between them? The coloring
of the edges was done just for such a purpose. You follow the
edges back from the end, K, to the first vertex. Thus the path
back to A is $K \to F \to D \to C \to A$, that is, the cheapest route is
$A \to C \to D \to F \to K$. (You can check that the cost of this path
is $\$199 + \$79 + \$99 + \$105 = \$402$.)

We point out that if you start at a different vertex, then you
will get different labels at the vertices—remember, the lengths
of the shortest paths are always from a fixed starting point, so if
you vary the initial point, you change the lengths.

For undirected graphs, you use Dijkstra's algorithm, either
by simply going along edges from discovered to undiscovered
vertices, or by replacing each undirected edge by two directed
edges, one in each direction, both labeled with the original
edge's weight (again a perfect example of "green" mathematics—
reusing a previously defined procedure for other purposes).
Here is the result of Dijkstra's algorithm on the graph of the
road map in Figure 3.28, starting from the upper left-hand
intersection. We see that the shortest distance from the upper left
corner to the upper right is 2121 feet, or just about four-tenths of
a mile.

In a **connected graph** (where there are walks between any
pair of vertices), every vertex gets a finite label, while in a

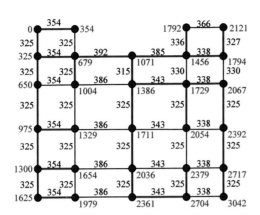

Figure 3.31: Shortest paths and distances from the upper left-hand corner via Dijkstra's algorithm.

disconnected graph, at least one label remains at ∞, so that Dijkstra's algorithm is one way to check whether a graph is connected or not. In fact, Dijkstra's algorithm also produces a spanning tree in the end, where the paths are the shortest paths from the root (where you start) to all other vertices.

We return, as promised, to mazes. What do shortest paths have to do with mazes, you ask? Well, let's consider a particular maze, as shown in Figure 3.32.

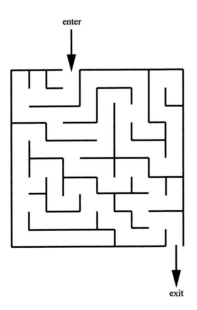

Figure 3.32: A maze.

This maze, like all others, consists of corridors, or hallways, that meet at openings; Figure 3.33 places a dot in each of the openings, including the start and finish (which open up to the outside of the room). The important observation about mazes is that *it doesn't matter how twisty and complicated the corridors are, all that is important, from the viewpoint of solving the maze, is which corridors meet at the openings.*

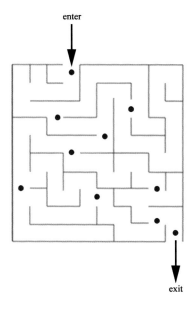

Figure 3.33: A maze with dots in the openings.

So we make a graph out of the maze by taking the openings as points, and joining two such vertices by an edge when a corridor directly joins them (see Figure 3.34). Corridors that are dead ends are of no use in solving the maze, so we ignore them.

Now all we seek is a path from the start vertex to the end vertex in the graph, as this will correspond of course to a solution to the maze. And we have an algorithm, Dijkstra's, to do such a thing. How do we fill in the edge weights to apply the algorithm? If we are just interested in any solution, we are free to choose the edge weights as we like—my personal choice would be to make all edge weights 1. If we do so, we can run the algorithm to get a solution (see Figure 3.35). If you want, you can measure the lengths of the corridors and use those lengths as your edge weights. Then the shortest path algorithm, starting

enter

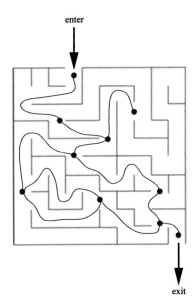

Figure 3.34: A maze's graph.

exit

at "enter" and ending at "exit", does indeed give you a shortest real-life path through the maze.

There are many, many other applications of shortest path algorithms. GPS (*Global Positioning Systems*), in conjunction with map programs, can help you find directions to a desired destination via shortest path algorithms on the underlying roadway graph, where the edge weights may be actual distance along a road or the expected time to traverse the road. Same algorithm, different edge weights!

And another fascinating application involves the *six degrees of separation principle*, namely, that in the social network of the world, there is a path of length at most six between any two people. If you have the graph of a social network, you can use the shortest path algorithm to find the shortest path connecting two individuals, but the whole graph would be too big to work with, even if it were available. But there is a related pastime called **six degrees of Kevin Bacon**, where one looks for shortest paths between the actor Kevin Bacon and any other given actor, dead or alive (for example, Kevin Bacon was in *Footloose* with John Lithgow, who was in *Terms of Endearment* with Shirley MacLaine, who was in *The Secret Life of Walter Mitty* with Ben

Stiller, so there is a path of length 3 between Kevin Bacon and Ben Stiller—see Figure 3.36).

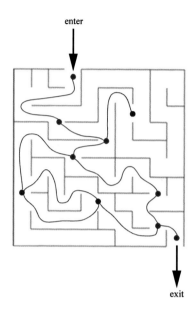

enter

exit

Figure 3.35: A maze's graph.

Back in 1994 Kevin Bacon opined that for any other actor, he had either worked with that actor or had worked with someone who had worked with that actor, and in light of the *six degrees of separation* principle, the *six degrees of Kevin Bacon* arose, and in fact there are **Bacon numbers**, which are the lengths of the shortest path to Kevin Bacon among actors (where the vertices of the graph are all Hollywood actors and the edges denote pairs that have worked together in a film). While most actors have low Bacon numbers, there are indeed some actors who (so far!) have Bacon numbers a little over 10.

According to researcher M. Habib, the **Kevin Bacon graph**, as of 2014, has more than 1.7 million vertices (i.e. actors) and more than 72 million edges. Of course, the graph changes on a regular basis, but at this time, there is an actor, Shemise Evans, whose shortest path to Kevin Bacon has length 8 (S. Evans → D. Bug → O. Karaoglu → F. Genckal → M. Ünal → A. Orak → A. Dawe → R. Serbedzija → K. Bacon), so that 6 degrees of Kevin Bacon only holds approximately!

Exercises

Exercise 3.2.1. Take as your vertices the positive integers $1, 2, 3, 4, 5$, with an edge between i and j if $i \neq j$ (that is, if they are different). Draw a picture of the graph.

Exercise 3.2.2. Take as your vertices the positive integers $1, 2, 3, 4, 5$, with an edge between i and j if i divides j evenly or j divides i evenly. Draw a picture of the graph.

Exercise 3.2.3. Take as your vertices the positive integers $1, 2, 3, 4, 5, 6, 7, 8, 9, 10$, with an edge

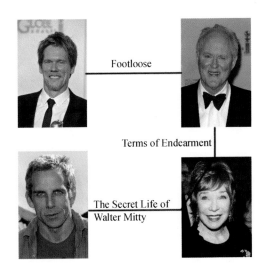

Figure 3.36: A path of length 3 in the Kevin Bacon graph.

Footloose

Terms of Endearment

The Secret Life of Walter Mitty

between i and j if i divides j evenly or j divides i evenly. Draw a picture of the graph.

Exercise 3.2.4. Suppose there are seven people in a class: Alice, Bob, Cindy, Doug, Ed, Francy and Greg. Here is the list of each person's friends:

- Alice's friends: Cindy, Doug, Francy

- Bob's friends: Doug, Francy, Greg

- Cindy's friends: Alice, Ed, Francy, Greg

- Doug's friends: Alice, Bob, Ed, Francy

- Ed's friends: Cindy, Doug, Francy, Greg

- Francy's friends: Alice, Bob, Cindy, Doug, Ed

- Greg's friends: Bob, Cindy, Ed

(a) Verify that the friendship relation is symmetric.
(b) Draw an undirected friendship graph for this group.
(c) Who has the most friends in the group? Explain what this corresponds to in the graph.
(d) Who have the fewest friends in the group? Explain what these corresponds to in the graph.
(e) Who would you say is the most *central* person in the group? Why?

Exercise 3.2.5. Now Alice, Bob, Cindy, Doug, Ed, Francy and Greg play off against one another in a badminton tournament—everyone plays everyone else exactly once. Here is a list of who beat whom:

- Alice defeated: Bob, Ed

- Bob defeated: Ed, Francy

- Cindy defeated: Alice, Bob, Francy, Greg

- Doug defeated: Alice, Bob, Cindy, Ed

- Ed defeated: Cindy, Greg

- Francy defeated: Alice, Doug, Ed, Greg

- Greg defeated: Alice, Bob, Doug

(a) Draw a directed graph whose arrows indicate who won each game.

(b) How many games were played? What does this correspond to in the graph?

(c) Was there a "winner" of the tournament? Explain your answer.

Exercise 3.2.6. Suppose that for the tournament described in Exercise 3.2.5, we say that person x is a "winner" if for every other person y, either x beats y or there is someone who x beats that defeated y. Under this definition, who would be declared a winner of the tournament?

Exercise 3.2.7. (Harder) A is a set of games where each person plays every other person exactly once, and there are no ties. Prove that in a tournament, there is always someone who is a "winner" under the definition proposed in Exercise 3.2.6.

Exercise 3.2.8. The following describes what each life form eats:

- crabs feed on: shrimp

- great white sharks feed on: sea turtles, stingrays

- stingrays feed on: crabs, shrimp, starfish

- sea turtles feed on: crabs, jellyfish

- shrimp feed on: jellyfish, phytoplankton

- starfish feed on: mussels

- jellyfish feed on: zooplankton

- mussels feed on: phytoplankton

- zooplankton feed on: phytoplankton

(a) Draw the corresponding food web.

(b) Redraw the graph of part (a) if necessary so that all the arrows point downwards. What property of the graph ensures that you can do so?

Exercise 3.2.9. The following shows an acquaintance graph.

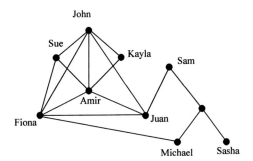

(a) Find a clique of four people. How many cliques of size four are there?

(b) What is the size of the biggest clique in the network?

(c) An **independent set** in an undirected graph is a set of vertices no two of which are adjacent. How would you describe in words what an independent set in an acquaintance graph corresponds to in real life?

(d) What is the largest independent set in the graph?

Exercise 3.2.10. For the acquaintance graph in Exercise 3.2.9:

(a) What is the length of the shortest path from John to Sasha?

(b) How many paths are there of the shortest length from John to Sasha?

Exercise 3.2.11. The length of a shortest path between vertices x and y in an undirected graph is called the **distance** between x and y, and written $\text{dist}(x,y)$ (if there is no path from x to y, we write $\text{dist}(x,y) = \infty$. Explain why the following hold:

(a) $\text{dist}(x,x) = 0$ for all vertices x.

(b) $\text{dist}(x,y) = \text{dist}(y,x)$ for all vertices x and y.

(c) $\text{dist}(x,z) = \text{dist}(x,y) + \text{dist}(y,z)$ for all vertices x, y and z.

(As expected, note that $\infty + a = \infty$ for all numbers a, and $\infty + \infty = \infty$.)

Exercise 3.2.12. Can you trace the following figure, without taking your pen off the page and without retracing a line? If so, give such a tracing, and if not, explain why not.

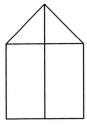

Exercise 3.2.13. Can you trace the following figure, without taking your pen off the page and without retracing a line? If so, give such a tracing, and if not, explain why not.

156

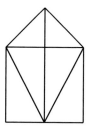

Exercise 3.2.14. Can you trace the following figure, without taking your pen off the page and without retracing a line? If so, give such a tracing, and if not, explain why not.

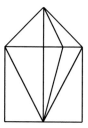

Exercise 3.2.15. Which of the graphs in Exercises 3.2.12, 3.2.13 and 3.2.14 can be traced starting and ending at the same point? Explain your answer.

Exercise 3.2.16. A **Hamiltonian cycle** is a cycle that goes through all of the vertices of a graph (exactly once). It may sound like the Eulerian circuit problem, but it appears to be much, much harder. Which of the following graphs have Hamiltonian cycles? Explain your answer.

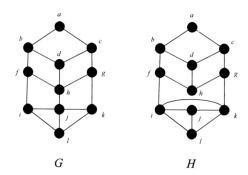

Exercise 3.2.17. After consulting another firm, the costs for running wires from office to office were updated as below. What is the minimum cost for connecting all of the offices?

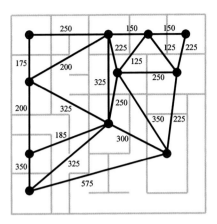

Exercise 3.2.18. For the weighted graph of Figure 3.28, what is the shortest distance from the upper left-hand corner to the upper right-hand corner?

Exercise 3.2.19. For the weighted graph of Figure 3.28, what is the shortest distance from the lower left-hand corner to the upper right-hand corner?

Exercise 3.2.20. For the weighted graph of Figure 3.29, find the cheapest flight from airport B to airport H.

Exercise 3.2.21. The **diameter** of a graph is the largest distance (see Exercise 3.2.11) between two different vertices. What is the diameter of the following graph?

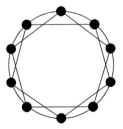

Exercise 3.2.22. How could you use Dijkstra's algorithm to find the diameter of a graph?

Exercise 3.2.23. Find your way out of the following maze, by using a graph.

158

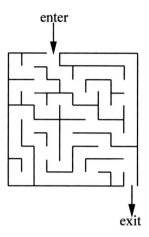

Exercise 3.2.24. Consider the following graph.

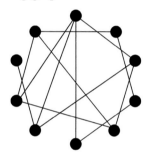

(a) How many edges does the graph have?

(b) What do you get if you add up all of the degrees of the vertices?

(c) Can you explain the relationship between parts (a) and (b) in general?

Exercise 3.2.25. What is the average degree of a vertex in the graph of Exercise 3.2.24? How can you calculate this in general from the number of vertices and edges?

Exercise 3.2.26. A **vertex k-coloring** (or sometimes just called a k-**coloring**) of a graph G is an assignment of one of k colors to each vertex so that adjacent vertices get different colors. For example, a 3-coloring of the graph

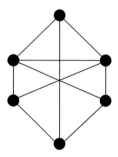

with the colors red, pink and black is shown below.

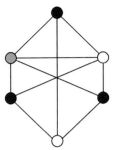

Find a 3-coloring of the graph of Exercise 3.2.24.

Exercise 3.2.27. The **chromatic number** of a graph is the smallest number k for which the graph has a k-coloring. What is the chromatic number of the graph in Exercise 3.2.21?

Exercise 3.2.28. The **eccentricity** of a vertex v in an undirected graph G is the maximum distance (see Exercise 3.2.11) of any vertex from v. For the following graph, find the eccentricity of vertex a. (Hint: use Dijsktra's algorithm for shortest paths, with all edge weights 1).

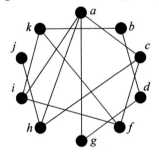

Exercise 3.2.29. For the graph of Exercise 3.2.28, find the eccentricities of all of the vertices.

Exercise 3.2.30. The **radius** of the graph is the smallest of all of the eccentricities. What is the radius of the graph of Exercise 3.2.28?

Exercise 3.2.31. The **path graph** P_n with n vertices is the graph whose vertices can be labeled as $v_1, v_2, \ldots, v_n$ with edges $v_1 v_2, v_2 v_3, \ldots, v_{n-1} v_n$. The following shows the path graph P_4 with 4 vertices.

(a) Which vertices of P_4 have the largest eccentricity?
(b) What is the diameter of P_4?
(c) What is the radius of P_4?
(d) Which vertices of P_n have the largest eccentricity?
(e) What is the diameter of P_n?

(f) What is the radius of P_n?

Exercise 3.2.32. The **cycle graph** C_n with n vertices is the graph formed from P_n by joining the first and last vertices of the path. The following shows the path graph C_4 with four vertices.

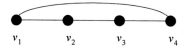

(a) Which vertices of C_4 have the largest eccentricity?
(b) What is the diameter of C_4?
(c) What is the radius of C_4?
(d) Which vertices of C_n have the largest eccentricity?
(e) What is the diameter of C_n?
(f) What is the radius of C_n?

Exercise 3.2.33. For the acquaintance graph in Exercise 3.2.9, find the eccentricities of all of the vertices and the radius. Based on these, who would you say are the most "central" people in the social network?

Exercise 3.2.34. Consider the graph below. Answer the following questions, explaining your answers.

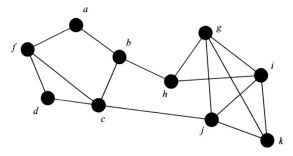

(a) What is the diameter of the graph?
(b) What is the radius of the graph?
(c) What is the size of a largest clique?
(d) What is the chromatic number of the graph?

Exercise 3.2.35. Suppose we have an undirected acquaintance graph. Explain why there must be two people who know exactly the same number of people, by converting the question into one about degrees in undirected graphs. (Hint: Use the **pigeonhole principle**, which states that if you have more objects than containers, and you place each object in a container, some container has to contain at least two objects.)

Exercise 3.2.36. A domino is a tile that contains a pair of numbers, from 0 (a blank) to 6. A complete set of dominoes consists of 28 tiles, one tile for each such pair. Can you arrange the tiles in a line so that adjacent tiles always share the same number? Explain your answer. (Hint:

Tiles are pairs of numbers, just as edges are pairs of vertices, so form a graph whose vertices are blank,1,2,3,4,5,6, joining each pair by an edge (corresponding to each domino). Then argue that what you are looking for is an Eulerian path, and one exists.)

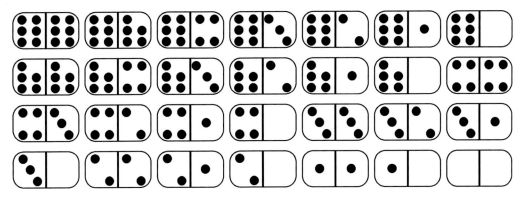

Exercise 3.2.37. A well known database for actors is IMDb, and is located at www.imdb.com. There you can enter an actor's name and explore the movies and television shows they have acted in (including the cast for each movie). Using this database, find a path of length 3 between Kevin Bacon and comedian/actor Robin Williams.

Exercise 3.2.38. Find the movies that corresponds to the path of length 8 from Shemise Evans to Kevin Bacon.

And Now the Rest of the Story . . .

Inspiration for a useful way to rank web pages arose like a surfer from the surf. Page and Brin imagined the web as a large directed graph, with the links as edges directed from one page to another. Under this visualization, they imagined a random web surfer, who would surf the web pages, via the links (that is, the directed edges), choosing a link at random from the web page he or she was on. The rank of a web page would be the proportion of time in the long run that the web surfer would spend on that page.

Figure 3.37: Larry and Sergey's company.

Now for certain, there were some mathematical challenges to be handled with this surfing model. For instance, how do we ensure that these rankings (which are really probabilities) could be always calculated? And even if they could be, how would you do it? For the web graph is HUGE (presently with billions of vertices and always growing), so whatever computation needs to be done, you'd have to be very clever about it to get it done in a reasonable amount of time.

But in the end, these were just details. Larry and Sergey finished off their work, dubbing their algorithm for web page rankings **PageRank**. *They implemented PageRank on the servers at Stanford University as the backbone to a clever search engine, and the program exceeded all expectations—it was so much better than any other such program available. The math was that good. The next step was to bring the research to the public domain via a company, which they called "Google," after the mathematical term "googol," which stands for 1 followed by one hundred 0's—a very big number indeed. And the rest, as they say, is history!*

3.3   *Review Exercises*

Exercise 3.3.1. For the following test scores (out of 10), draw a bar chart. What can you say by looking at the bar chart?

$$8, 3, 6, 2, 3, 7, 6, 6, 4, 5, 5, 4, 5, 9, 7.$$

Exercise 3.3.2. Suppose there was an election with three candidates: Alan, Bert and Carl. They respectively got 30%, 50% and 20% of the vote. Draw an appropriate pie chart.

Exercise 3.3.3. The following table shows the incidence of a viral disease from 2001 to 2014.

Year	Number of new cases
2001	5
2002	31
2003	58
2004	117
2005	498
2006	613
2007	792
2008	844
2009	705
2010	723

Draw a line graph for the data.

Exercise 3.3.4. List everything that you can find wrong with the following graphic.

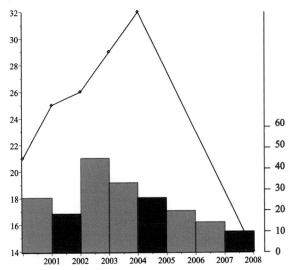

Exercise 3.3.5. Draw a food web graph with the following animals and plants: hawk, owl, squirrel, fox, mouse, grasshopper, robin, rabbit, spider, grass.

Exercise 3.3.6. Trace the following figure without taking your pen off the page and without retracing a line. Do you necessarily have to start and end at the same place?

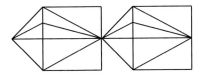

Exercise 3.3.7. After consulting another firm, the costs for running wires from office to office were updated as below. What is the minimum cost for connecting all of the offices?

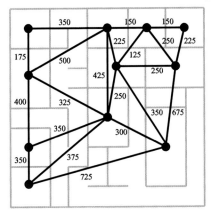

Exercise 3.3.8. For the weighted graph of Figure 3.28, what is the shortest distance from the lower left-hand corner to the lower right-hand corner?

Exercise 3.3.9. Suppose we have to schedule the final exams for 10 courses, each in a 3-hour time slot. Two courses conflict with one another if some student takes both courses, and we want to avoid scheduling two conflicting courses in the same time slot. Here is a list of the conflicting courses:

- MATH 1000 conflicts with MATH 2030 and MATH 2135

- MATH 1500 conflicts with MATH 2135, MATH 3030 and MATH 3500

- MATH 1800 does not conflict with any course

- MATH 2030 conflicts with MATH 1000, MATH 3030 and MATH 4460

- MATH 2135 conflicts with MATH 1000, MATH 1500, MATH 4460 and MATH 4900

- MATH 3030 conflicts with MATH 1500, MATH 2030, MATH 4460, MATH 4800 and MATH 4900

- MATH 3500 conflicts with MATH 1500 and MATH 4900

- MATH 4460 conflicts with MATH 2300, MATH 2135, MATH 3030 and MATH 4800

- MATH 4800 conflicts with MATH 3030, MATH 4460 and MATH 4900

- MATH 4900 conflicts with MATH 2135, MATH 3030, MATH 3500 and MATH 4800

(a) Draw a graph whose vertices are the courses and whose edges denote courses that conflict.
(b) Find a schedule for the final exams with the fewest number of 3-hour time slots. (Hint: think of the time slots as different colors!)

Exercise 3.3.10. Find a path of length at most 4 between Wanda Sykes (the comedian) and Buzz Aldrin (the astronaut!) in the IMDb graph.

Money and Risk

It was a warm Friday evening on July 5, 2013. The engineer, Tom Harding, was driving the train down the tracks in rural Quebec, in Eastern Canada. Of the 79 freight train cars, 5 were locomotives, at the head of the train, while the back end consisted of 72 tank cars, all carrying crude oil. One of the other cars was used as a buffer between the front and back ends, and the final one, a so-called "VB" car, contained equipment for the one-man operation of the train.

You see, in recent years, the Montreal, Maine and Atlantic Railway (MMA), the owner of the train and rail line, had been trying to cut costs wherever possible and was permitted by both American and Canadian governments to have freight trains run under the SPTO (Single Person Train Operation) protocol. Just one man on freight trains that averaged 80 cars. Lots of lonely miles and lots of responsibility for just one person.

The scenery, as always, was soothing. Quaint. Quiet. Beautiful. It was after 10:00 P.M., late by any standards, and Tom, one of the firm's best engineers, decided to call it a night in Nantes, about 7 miles out of the nearby town of Lac-Mégantic. He would have preferred to have parked the train on the siding nearby, rather than right on the main tracks, but the siding was full of empty boxcars. He knew he had to leave one of the locomotives running to keep air pressure high enough for the air brakes. After setting the air brakes throughout and setting manual brakes on the locomotives and 10 of the 72 oil tankers, Harding left by taxi for one of the hotels in Lac-Mégantic. The train sat idling on the main tracks, no one on board, unlocked.

Around 11:30 a passerby noticed a fire on board the first locomotive and called 911. The Nantes firefighters responded very quickly, shutting off the engine to facilitate putting out the fire. After informing MMA of the fire, they confirmed with two train employees that the train was safe. Everyone left a little after midnight.

Nantes slopes slightly down toward Lac-Mégantic. Close to 1:00 A.M. the train started rolling down toward the sleepy town, slowly at first, with no lights on. But over the 7 miles, the train gathered speed.

Townsfolk sipping coffee at the Musi-Café, in the town's center, had to rub their eyes in disbelief as a massive dark object barreled toward them, with the earthquake-like rumble shaking them to the core.

Nothing fills our brains like decision making. We take in whatever data we can and run all sorts of mental processes on it to predict the best outcome. Sometimes the desired outcome is to maximize the money we get, sometimes to minimize the money we lose, and sometimes it involves non-monetary issues, like being safe, having fun, getting love, and achieving a goal. It's surprising, though, how little we think about the process of decision making. Is there a scientific way to make better decisions? We could all benefit from that. And indeed mathematics can inform the way we make decisions.

4.1   Money—Now or Later

TIME IS MONEY, SAID BENJAMIN FRANKLIN. But it is also true that money is time—a dollar now is worth more in the future, and a future dollar is worth less now. As money percolates through most of our daily decisions, it's worthwhile understanding just exactly how money and time work together.

A Healthy Interest in Interest

The key to understanding how money and time work together involves the notion of **interest**, the amount you can earn by investing money over time. The **interest rate** is the *percentage* of the original amount invested (called the **principal**) that you earn over a certain time period. For example, with an interest rate of 10% per year (that is, annually, or, using archaic Latin, *per annum*), at the end of one year of investing a principal of $2500, you would earn "10% of $2500," that is (using our translations of Chapter 1),

$$0.10 \times \$2500 = \$250.$$

If this was the end of the investing period, you would remove your principal, $2500, and collect the interest, $250 as well. You can think of the interest as the money you earned by allowing the borrower the privilege of using the principal for the year.

Calculating interest is pretty easy if the stated period for the interest rate and the period of investing (or borrowing) is the same:

If you invest P dollars at a rate of i (a percentage, written as a decimal) per period of time, then, at the end of the period,

$$\text{interest earned} = Pi$$

and

$$\text{total amount} = P + Pi = P(1 + i).$$

Now rarely are things this simple! Sometimes the rate is given for one period, say a year, but interest is calculated on a shorter period. For example, if you are quoted a yearly rate of 9%, then the monthly rate is calculated by dividing the yearly rate by 12, as there are 12 months in a year. So the monthly rate is $\frac{9\%}{12} = 0.75\%$, or 0.0075, written as a decimal (they often call the yearly stated rate the **nominal interest rate**, and it is used to set the rate per period by dividing). If the interest is calculated daily, then the daily rate would be $\frac{9\%}{365} = 0.0247\%$, or 0.000247 as a decimal. A small number to be sure, but if the principal were 1 000 000, then the interest over one day would be $0.000247 \times \$1\,000\,000 = \$246.58$—not exactly small to me.

And people like to earn more money by reinvesting. There are two ways that interest is paid over longer periods of time— **simple interest**, where interest is only earned on the original principal (and you are free to take the interest out after each interest payment, if you wish), and **compound interest**, where interest is reinvested with the principal, and interest is earned on all. For example, suppose on a principal of $10 000 you earned 12% yearly interest, payable monthly (each month, the interest is therefore $\frac{12\%}{12} = 1\% = 0.01$). With simple interest, each month

Nominal mean "in name only," as the rate stated is only used for calculating the rate for smaller periods, such as monthly or daily.

you earn

$$0.01 \times \$10\,000 = \$100$$

so over the whole year, you would earn this amount every month, for a total of $12 \times \$100 = \$1200$ in interest, and receive back principal plus interest, that is, $\$10\,000 + \$1200 = \$11\,200$, at the end of the year.

Things are different (and much better!) with compound interest. At the end of the first month, you earn $0.01 \times \$10\,000 = \$100$, just as with simple interest. But for the second month, you earn interest not just on the principal, $\$10\,000$, but on the principal plus interest, $\$10\,000 + \$100 = \$10\,100$, so you earn 1% of this amount, that is

$$0.01 \times \$10\,100 = \$101.$$

That's only a difference of one dollar over the $\$100$ you would earn with simple interest for the second month, but the advantage lies over longer periods of time. The next month, you earn 1% of $\$10\,100 + \$101 = \$10\,201$, which is $\$102.01$. This takes you to $10\,303.01$, compared to the $\$10\,300$ you would have after three months of simple interest—a whole $\$3.01$ more. Sounds like much ado about nothing.

Think again. Note that the amount present after one month is

$$
\begin{aligned}
\$10\,000 + 0.01 \times \$10\,000 &= \$10\,000 \times 1 + 0.01 \times \$10\,000 \\
&= \$10\,000 \times (1 + 0.01) \\
&= \$10\,000 \times 1.01,
\end{aligned}
$$

by factoring. After two months, the amount is the amount after one month, $\$10\,000 \times 1.01$, plus itself times 0.01, that is,

$$
\begin{aligned}
&\$10\,000 \times 1.01 + 0.01 \times (\$10\,000 \times 1.01) \\
&= (\$10\,000 \times 1.01) \times 1.01 \\
&= \$1000 \times 1.01^2.
\end{aligned}
$$

The pattern continues—after 3 months, the amount in the account is $\$10\,000 \times 1.01^3$, and so on. At one year, that is, after 12 months, what is in the account is

$$\$10\,000 \times 1.01^{12} = \$11\,268.25,$$

which is $68.25 more than the $11 200 you would have in the account with simple interest.

The results are more spectacular over even longer periods of time. Table 4.1 shows what would be in the account after 2, 3, 5 and 10 years of simple and compound interest, respectively. Over 10 years, the gap is huge—you'd have over $11 000 more in the compound interest account than in the simple interest one. It certainly pays to reinvest the interest!

Number of years	Simple Interest	Compound Interest
2	$12 400	$12 697.35
3	$13 600	$14 307.69
5	$16 000	$18 166.97
10	$22 000	$33 003.87

Table 4.1: Comparison of totals under simple and compound interest at 1%/month on a principal of $10 000.

The formulas for simple interest and compound interest come up so often that they are worth remembering. For simple interest, if the principal is P dollars and the interest rate is r per period, then over t periods, the interest per period is Pi, so over the t periods you earn $Pi \times t = Pit$, giving a total amount in the account of $P + Pit = P(1 + it)$. For compound interest, the amount in the account is $P(1 + i)^t$. We summarize:

If you invest P dollars at a rate of i (a percentage, written as a decimal) per period, then, at the end of t periods, in a simple interest account you would have

$$A = P(1 + it) \text{ dollars}$$

while in a compound interest account you would have

$$A = P(1 + i)^t \text{ dollars.}$$

We point out that when you are dealing with compound interest where the interest is paid at a period other than a year, there is a notion of an **effective annual rate** of interest which differs from the nominal annual interest rate. The effective annual rate is the annual rate that is equivalent to that of the period's interest rate, when you compound money for the whole year. For example, when the interest rate is 1% per month, then under compound interest, the dollar would accumulate to $1 × 1.01^{12} = $1.1268, so that the interest over the year would be 0.1268, which is 12.68%, and that is the effective annual interest rate, compared to the nominal interest of 12% (calculated by multiplying the monthly interest rate, 1%, by 12, the number of months in a year). For any amount of money, you would earn the same amount of interest over a year whether you invest it at a nominal rate of 12%, compounded monthly, or at a yearly rate of 12.68%. The effective annual interest rate is always bigger than the nominal annual interest rate. It may look like a small difference here, but when you have lots of money invested, it can make a *big* difference.

Back in Chapter 1 we talked about functions. If the principal P and rate i are fixed, and we view the account totals (principal plus interest), A, as a function of the number of time periods, then

- under simple interest, $A = P + (Pi)t$, which is the equation of a straight line with vertical intercept P and slope Pi, while

- under compound interest, $A = P(1 + i)^t$, which is the equation of an *exponential function*, which grows large fairly quickly.

Figure 4.1 illustrates the difference with $P = 10\,000$ and $i = 0.01$. So, given a choice (which is unlikely, but nevertheless!):

Life Lesson: For a fixed principal and interest rate, it is always better to have compound interest over simple interest if you are investing, and vice versa if you are borrowing!

Investing in Your Future

This brings us back to the value of money now versus money in the future. What should the value of a certain amount of money now be worth in the future? The most natural way to calculate this is by what the money, if invested in an account now, at the current interest rate, would be worth at a point in the future. This of course depends on the principal, the interest rate, the time period, and the method of calculating interest. But given these, we have a way to calculate the value of future money. So, for example, the future value of $100 000, in 10 years, at a yearly interest rate of 9%, is

- $100 000(1 + 0.09 \times 10) = \$190\,000$ under simple interest, and

- $100\,000(1 + 0.09)^{10} = \$236\,736.37$ under compound interest.

Again, future money is worth more than present money, and more under compound than simple interest.

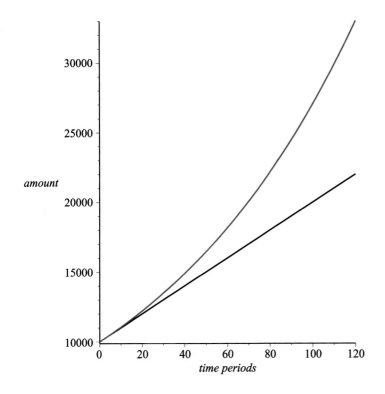

Figure 4.1: Simple (red) versus compound (black) interest.

Back to the Present

We can turn things on their ears and ask: what is the present value of future money? What is the value of $10 000 10 years from now, if the interest rate is 9% per year? Here a mathematical viewpoint makes the answer clear—the **present value** should be the principal P we should invest right now to leave us with $A = \$100 000$ 10 years from now. Under simple interest, we have seen that $A = P(1 + it)$, so with $A = 100 000$, $i = 0.09$ and $t = 10$, we find that the present value P must satisfy

$$100 000 = P(1 + 0.09(10)),$$

so, solving for P, we get

$$P = \frac{100 000}{1 + 0.09(10)} = 52 631.58.$$

As a check, if we invest $52 631.58 now for 10 years at 9% interest per annum, with simple interest, we will have, at the end of 10 years,

$$A = P(1 + rt) = 52 631.58(1 + 0.09(10)) = 52 631.58 \times 1.9 = 100 000,$$

as desired. Thus the present value of $100 000, under this scenario, is $52 631.58.

What about under compound interest? The math is pretty much the same, only a bit more complicated. In this case, we have that $A = P(1 + i)^t$, so that with $A = 100 000$, $i = 0.09$ and $t = 10$, we have

$$100 000 = P(1 + 0.09)^{10},$$

so, solving for P, we get

$$P = \frac{100 000}{1.09^{10}} = 42 241.08.$$

Checking once again, if we invest $42 241.08 now for 10 years at 9% interest per annum, with compound interest, we will have, at the end of 10 years,

$$A = P(1 + r)^t = 42 241.08(1 + 0.09)^{10} = 52 631.58 \times 1.9 = 100 000,$$

as desired. Thus the present value of $100 000, under this scenario, is $42 241.08. So given a choice between receiving $100 000

It may seem surprising that under compound interest the present value is smaller than under simple interest, but it does make sense upon second thought—after all, the present value is the amount you need now to have a fixed amount in the future, and as compound interest yields more than simple interest, you need less to start with under compound interest, compared to simple interest, to reach the same amount.

10 years from now and say $35 000 right now, you should opt for the former, as the present value of $100 000, 10 years from now, is $42 241.08, more than the $35 000 you are being offered.

When working with present and future values, you can always work from first principles (as opposed to first *principals*!), remembering how interest is calculated, and I recommend that you do so as often as you can. But for those who like formulas:

The present value of A dollars, due in t periods, at a rate of i (a percentage, written as a decimal) per period, under simple interest, is

$$P = \frac{A}{1 + it} \text{ dollars}$$

while under compound interest, it is

$$P = \frac{A}{(1 + i)^t} \text{ dollars.}$$

Making Installments

In addition to making a straight-up investment for a period of time or taking out a one-time loan with a single payment at the end of the loan, what often happens in practice is that either you set aside money each time period and put it into an investment account to grow, or you repay your loan with installment payments. Both of these are really the same thing— all that differs is who pays and who gets paid!

So let's do an example. Suppose that you decide that you will set aside $100 every month, for the next 5 years, deposited into an account that pays 9% interest, compounded monthly. Figure 4.2 shows the timeline for the deposits. There are $12 \times 5 = 60$ deposits, each of $100, over the 5 years. Without any interest, there would be $61 \times \$100 = \$6100$ in the account after 5 years (there are not 60 but 61 deposits as you deposit at the beginning of each month, including the first and the sixty-first, at the end). What does the compound interest add?

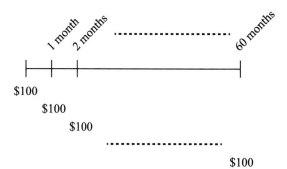

The key to realize is that each deposit independently earns
compound interest at a rate of $\dfrac{9\%}{12} = 0.75\% = 0.0075$ over the
number of months that it is invested. The following timeline
shows how much each deposit accumulates in the account:

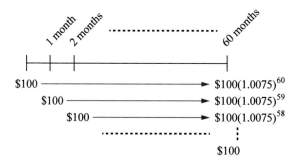

Again, a picture is worth a thousand words! All we have to
do to find the total in the account after 5 years is add up the
following dollar amount:

$$100(1.0075)^{60} + 100(1.0075)^{59} + 100(1.0075)^{58} + \cdots + 100$$
$$= \quad 100 + 100(1.0075) + 100(1.0075)^2 + \cdots + 100(1.0075)^{60}$$

(All I've done is reorder the terms.) Of course, you can use a
calculator now to add this up, but that is a lot of work—there
are 61 terms to work out and add together. This **series** (which
is simply the word for a sum of terms) is of a special type—a
geometric series, where, for some number r, the next term to be
added is r times the previous term (which is the same as saying
that the ratio of any one term divided by the previous one is r).

In this example, $r = 1.0075$. So suppose, in general, we want to add up a geometric series

$$S = a + ar + ar^2 + \cdots + ar^n.$$

Here a is the first term, and r is what we call the **common ratio**. There is a formula, but rather than just give that to you, let's figure it out! What we do is multiply the sum by r, and use a bit of arithmetic by distributing r through the bracket:

$$\begin{aligned} Sr &= (a + ar + ar^2 + \cdots + ar^n)r \\ &= ar + ar^2 ar^3 + + \cdots + ar^{n+1} \end{aligned}$$

Now the brilliant trick is to recognize that most of the terms in the series for S and Sr are the same and therefore will cancel if we line things up properly and subtract:

$$\begin{array}{rcllllllll} S &=& a &+& \cancel{ar} &+& \cancel{ar^2} &+& \cdots &+& \cancel{ar^n} \\ Sr &=& & & \cancel{ar} &+& \cancel{ar^2} &+& \cdots &+& \cancel{ar^n} &+& ar^{n+1} \\ \hline S - Sr &=& a & & & & & & & & & - & ar^{n+1} \end{array}$$

By factoring out S on the left side, we see that $S(1 - r) = a - ar^{n+1}$, and by dividing the left side by $1 - r$ (provided it isn't equal to 0, that is, provided r isn't 1), we find that

$$S = (a - ar^{n+1})/(1 - r),$$

that is, S is the first term minus the term *after* the last one, all divided by 1 minus the common ratio. This is so useful we'll put it in a box for reference:

Summing a geometric series: Suppose a and r are any numbers. The geometric series

$$a + ar + ar^2 + \cdots + ar^n$$

has sum

$$\frac{a - ar^{n+1}}{1 - r}$$

provided r isn't 1 (if $r = 1$, then the sum is obviously $(n+1)a$ as all terms are equal to a).

Returning to our example of the deposits of $100 at the beginning of every month for 6 years at 9% nominal annual interest compounded monthly, we need to add up

$$100 + 100(1.0075) + 100(1.0075)^2 + \cdots + 100(1.0075)^{60}.$$

This is a geometric series with first term $a = 100$, common ratio $r = 1.0075$ and $n = 60$, so the sum, from the formula just noted, is

$$\frac{100 - 100 \cdot 1.0075^{61}}{1 - 1.0075} = \frac{100 - 157.74}{-0.0075} = \frac{-157.74}{-0.0075} = 7698.67.$$

That is, you would have $7698.67 in your account, $1598.67 more than you would have had just from the 61 deposits of $100.

Alternatively, if you don't remember the formula, but just the process for summing a geometric series, you can find the sum, which we'll call S, by

$$
\begin{array}{lclcl}
S & = & 100 & + & \cancel{100(1.0075)} & + & \cdots & & & + & \cancel{100(1.0075^{60})} \\
S(1.0075) & = & \cancel{100(1.0075)} & + & \cdots & & & + & \cancel{100(1.0075^{60})} & + & 100(1.0075^{61}) \\
\hline
S - S(1.0075) & = & 100 & & & - & 100(1.0075^{61})
\end{array}
$$

so $-0.0075S = 100 - 100(1.0075^{61})$, that is, with a bit of algebra,

$$S = \frac{100 - 100(1.0075)^{61}}{-0.0075} = 7698.67$$

again. Is it better to use the formula or do it out by hand? It's a matter of choice and comfort, but I like to bypass the formula and practice the math.

The process is the same for loans with equal payments, such as mortgages. For example, suppose you take out a $350 000 mortgage on a house, over 30 years, at a nominal annual interest rate of 6%, compounded monthly (so the monthly rate is $0.06/12 = 0.005$). The money you receive is not a future amount, to be paid to you 30 years down the road, but present money, to be given right away. You are to pay K dollars at the end of each month. How does the bank know what to charge for K?

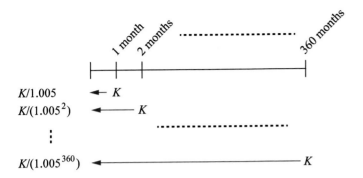

Forming a timeline again, but pulling the payments back to the present, we need the value right at the onset (the *present value*) of all of the payments to be equal to the amount you borrow, \$350 000. That is, our **equation of value**, which equates at one point in time (here the start) the values of the amounts received versus the amounts paid, is

$$350\,000 = \frac{K}{1.005} + \frac{K}{1.005^2} + \cdots + \frac{K}{1.005^{360}}$$

$$= \frac{K}{1.005} + \frac{K}{1.005}\left(\frac{1}{1.005}\right) + \frac{K}{1.005}\left(\frac{1}{1.005}\right)^{359}.$$

I purposely rewrote the right side so that we can see that each term on the right is $\frac{1}{1.005}$ times the previous one, that is, the right side is a geometric series with first term

$$a = \frac{K}{1.005},$$

common ratio

$$r = \frac{1}{1.005},$$

and

$$n = 359.$$

We don't need any new formula—we can reuse and recycle the one we have at hand, that for a geometric series. All that changes are the parameters a, r and n. This is as green as it gets. Thus the right side of our equation of value is

$$\frac{\frac{K}{1.005} - \frac{K}{1.005}\left(\frac{1}{1.005}\right)^{360}}{1 - \frac{1}{1.005}}$$

which, with a little algebra, which I leave to you, can be simplified to

$$K \left(\frac{1 - \left(\frac{1}{1.005} \right)^{360}}{0.005} \right).$$

Therefore, our equation of value now looks like

$$350\,000 = K \left(\frac{1 - \left(\frac{1}{1.005} \right)^{360}}{0.005} \right),$$

and solving for the monthly payment, K, we get

$$K = \frac{350\,000 \times 0.005}{1 - \left(\frac{1}{1.005} \right)^{360}} = 2098.427$$

which we'll round to the nearest cent, so our monthly payment is \$2098.43. And that is what the mortgage company will charge you. What you end up paying over the 30 years in total is $360 \times \$2098.43 = \$755\,434.80$)—yikes! A huge return for the bank on the mortgage and a depressing thought for you as the borrower. And the way it is set out, you pay mostly interest at the beginning, with very little repayment of the principal, until toward the end.

Is there a formula for the calculation? For sure, and if you were going to work in the financial industry, it would be worthwhile to memorize it. But here we are more interested in the approach—determining an equation of value and using only the formula for geometric series—so I'll forego giving the formula. An idea in the head is worth more than two formulas in a book, I always say!

Annuities—Money in the Bank, Now and Forever?

Now in the mortgage example (which is a loan), you borrow money from the financial institution and make equal repayments. But you could turn this around, with you lending the financial institution money and it paying you equal payments until their loan to you is paid off. This is called an **annuity** and the math is just the same. So if you loaned a bank \$350\,000 for 30 years at a nominal annual rate of 6% compounded monthly, then you would be entitled to receive equal monthly payments of $2048.43 over the 30 years.

Most often, for an annuity, rather than invest a certain amount and then figure out what the monthly payment to you should be, what you often want is to guarantee a certain monthly payment, and then determine how much to invest right now. Let's do an example that's a bit complicated. Suppose that, planning for your early retirement, you turned 25 today and you want to receive yearly payments of $15 000 for 30 years, starting one year after turning 50. The interest rate is set at 8% per annum (compounded yearly). How much should you invest? The money will sit in the account for 25 years, and then you will start receiving your annuity, at the end of each year, for 30 years. A timeline for the payments is as follows (each payment is shown at the end of the year, just as the new year turns):

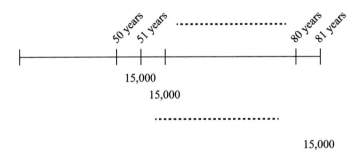

We pull back each of these payments to the present, using the present value formula:

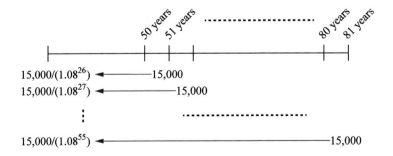

The total of these is what you need to invest now for the annuity. We'll use a bit of algebra and again our very useful geometric series formula (once we see the geometric series):

$$
\begin{aligned}
\text{present investment} \quad &= \quad \frac{1500}{1.08^{26}} + \frac{15\,000}{1.08^{27}} + \cdots + \frac{15\,000}{1.08^{55}} \\[2mm]
&= \quad \frac{15\,000}{1.08^{26}}\left(1 + \frac{1}{1.08} + \left(\frac{1}{1.08}\right)^2 + \cdots + \left(\frac{1}{1.08}\right)^{29}\right) \\[2mm]
&= \quad \frac{15\,000}{1.08^{26}}\left(\frac{1 - 1 \times \left(\frac{1}{1.08}\right)^{30}}{1 - \frac{1}{1.08}}\right) \\[2mm]
&= \quad 24\,657.57
\end{aligned}
$$

So provided you can afford to invest right now, only a relatively small amount is needed to supply you with a yearly annuity of $15\,000.

Now this is fine if you only plan to live for 80 years, but what if you live longer? How long? Perhaps medicine will advance so much by then that the sky is the limit in terms of longevity. So what about if you want payments to continue indefinitely? Is such a thing possible? Indeed it is, and it is called a **perpetuity**. It might seem impossible—how can you set aside enough money now for guaranteed recurrent payments *forever*—but the math says you can, and takes us into a fascinating topic, **infinite series**.

Let's do an example. Suppose we want to invest enough money right now so that starting at the end of each year, we'll receive $1000, *forever*. The interest rate is 8% per annum. The timeline looks as follows:

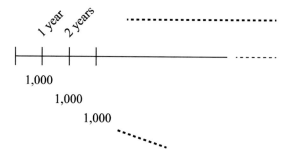

Again, to find the money we need to invest *now*, we add up the present values of all of the payments:

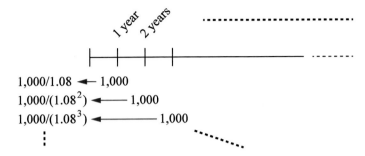

1,000/1.08 ◄── 1,000
1,000/(1.08²) ◄──── 1,000
1,000/(1.08³) ◄───── 1,000
⋮

The problem is that we need to add up infinitely many terms. But suppose we, in our minds, stop the payments after some large number of them, say 100. Then we find that the amount we need to invest now is

$$\frac{1000}{1.08} + \frac{1000}{1.08^2} + \cdots + \frac{1000}{1.08^{100}}$$

$$= \frac{1000}{1.08}\left(1 + \frac{1}{1.08} + \left(\frac{1}{1.08}\right)^2 + \cdots + \left(\frac{1}{1.08}\right)^{99}\right)$$

$$= \frac{1000}{1.08}\frac{1 - \left(\frac{1}{1.08}\right)^{100}}{1 - 1/1.08}$$

$$= 12\,494.32$$

So that gives us a rough idea of how much we need to invest now. But of course we need more (if we can do it at all) as we want not just 100 payments, but an indefinite number of payments. If we do the same thing for 1000 payments, we get

$$\frac{1000}{1.08} + \frac{1000}{1.08^2} + \cdots + \frac{1000}{1.08^{1000}}$$

$$= \frac{1000}{1.08}\frac{1 - \left(\frac{1}{1.08}\right)^{1000}}{1 - 1/1.08}$$

$$= 12\,500.00,$$

though there is some rounding going on. But the key thing is that things didn't change very much when going from 100 to 1000 payments.

Why? We can see this from seeing what changed in the two calculations—we just replaced $\left(\dfrac{1}{1.08}\right)^{100}$ by $\left(\dfrac{1}{1.08}\right)^{1000}$. But these are both really, really small, as we are raising the number $1/1.08 = 0.9259$, which is strictly between -1 and 1, to higher and higher powers, and this goes to 0. So we see that as we increase the number of payments we receive more and more, the present values of the payments get closer and closer to

$$\frac{1000}{1.08}\frac{1-0}{1-1/1.08} = \frac{1000}{0.08} = 12\,500$$

exactly! This we take as the value of the infinite series—what the bigger and bigger **partial sums** approach—*and* the value we need to invest now for the perpetuity. Have we done things correctly? Does it really all make sense? Note that after one year of investing $12\,500$ at 8% per annum, our interest on the principal is $0.08 \times \$12\,500 = \$1000$, so scoping this out as our yearly payment, we still have the original principal, $\$12\,500$, going forward, and indeed we can take out $\$1000$ every year, for the same reason, forever!

Summing infinite series leads us into a field of mathematics called calculus, where words like "approach" are made mathematically precise, but this is as far as we'll go here. We'll summarize the mathematics of adding up an infinite geometric series:

Summing an infinite geometric series: Suppose a and r are any numbers, with r strictly between -1 and 1. The infinite geometric series

$$a + ar + ar^2 + \cdots + ar^n + \cdots$$

has sum

$$\frac{a}{1-r}.$$

(If r is outside this range, that is, $r \le -1$ or $r \ge 1$, and a isn't 0, then the infinite geometric series doesn't add to a value.)

Whether just for a short time, for a lifetime, or forever, understanding the time value of money is crucial to making good monetary decisions.

Exercises

Exercise 4.1.1. If the stated nominal interest rate is 12%, what is the monthly rate? The weekly rate? The daily rate?

Exercise 4.1.2. If the monthly rate is 0.75%, what is the stated nominal annual interest rate? The weekly rate? The daily rate?

Exercise 4.1.3. You read online that the interest rate being offered by a financial institution is 5% per annum.

(a) If you are taking out a loan, do you prefer the interest to be calculated as simple or compound?

(b) If you are depositing the money into a savings account, do you prefer the interest to be calculated as simple or compound?

Exercise 4.1.4. You are ready to invest $1200. You have a choice of two accounts: Account A pays simple interest at a rate of 9% per year and Account B pays compound interest at a rate of 9% per year, as well. Complete the following table for the amount in each account at the end of each year, just after interest has been paid.

Year	Account A	Account B
1		
2		
3		
4		
5		

Exercise 4.1.5. You are ready to invest $10 000. You have a choice of two accounts: Account A pays simple interest at a rate of 6% per year and Account B pays compound interest at a rate of 6% per year, as well. Complete the following table for the amount in each account at the end of each year, just after interest has been paid.

Year	Account A	Account B
10		
20		
30		
40		
50		

Exercise 4.1.6. You are ready to invest $1000. You have a choice of two accounts: Account A pays simple interest at a rate of 8% per year and Account B pays compound interest at a rate of 5% per year. Complete the following table for the amount in each account at the end of each year, just after interest has been paid.

Year	Account A	Account B
1		
2		
3		
4		
5		
10		
20		

Exercise 4.1.7. Suppose you invest $2000 for one year at an interest rate of 10% per annum. After one year, how much interest will you get? What will be the total amount in the account then? Does it matter whether the interest is simple or compound?

Exercise 4.1.8. Suppose you invest $3500 for three years at a simple interest rate of 8% per annum. After one year, how much interest will you get? What will be the total amount in the account at the end of three years?

Exercise 4.1.9. Suppose you invest $3500 for three years at an interest rate of 8% per annum, compounded yearly. After one year, how much interest will you get? What will be the total amount in the account at the end of three years?

Exercise 4.1.10. Suppose you invest $10 000 and after four years you have $12 800. What interest did you earn? If the interest was simple interest, what was the interest rate?

Exercise 4.1.11. You take out a loan of $15 000 at 6% per annum simple interest. If you don't make any payments until the end of the fifth year and then pay it back, how much will you return to the lender?

Exercise 4.1.12. For Exercise 4.1.11, suppose the compound interest is calculated at 0.5% per month. Does the answer change? Why?

Exercise 4.1.13. You take out a loan of $15 000 at 6% per annum interest, compounded annually. If you don't make any payments until the end of the fifth year, and then pay it back, how much will you return to the lender?

Exercise 4.1.14. For Exercise 4.1.13, suppose the simple interest is calculated at 0.5% per month. Does the answer change? Why?

Exercise 4.1.15. Suppose you invest $3000 at a nominal 5% annual rate, compounded monthly. What is the effective annual rate of interest?

Exercise 4.1.16. Suppose you invest $3000 at a nominal 6% annual rate, compounded monthly. What is the effective annual rate of interest?

Exercise 4.1.17. What is the present value of $4000, five years in the future, if simple interest is calculated at a rate of 7% per annum? To what does this present value accumulate in five years, with simple interest at 7% per annum?

Exercise 4.1.18. What is the present value of $4000, five years in the future, if compound interest is calculated at a rate of 7% per annum? To what does this present value accumulate in five years, with simple interest at 7% per annum?

For Exercises 4.1.19 to 4.1.28, draw out a timeline to get an equation of value at some point in time for all payments, and use the latter to solve for the unknown amount.

Exercise 4.1.19. You put $2000 in an account that pays 8% annual interest, compounded annually—this is the start of year one. You then put in another $1000 at the beginning of year three and $1500 at the beginning of year five. How much is in the account at the beginning of year nine?

Exercise 4.1.20. You put $1000 in an account that pays 6% annual interest, compounded annually—this is the start of year one. You then put in another $500 at the beginning of each of years two through eight. How much is in the account at the beginning of year nine?

Exercise 4.1.21. You put $10 000 in an account that pays 5% annual interest, compounded annually—this is the start of year one. You then put in another $5000 at the beginning of year three. You want to have $24 000 in the account at the beginning of year nine. How much should you put into the account at the beginning of year five for this to happen?

Exercise 4.1.22. Suppose you took out a loan of $5000 a year ago, at 6%, compounded monthly—you are required to repay the loan in full at the end of three more years. Interest rates have climbed to 9%, and you have decided to deposit enough money in an account that pays 9% annual interest, compounded monthly, to pay out the loan when it is due. How much money do you need to deposit into the account now?

Exercise 4.1.23. You take out a mortgage of $225 000 over 30 years at a nominal annual interest rate of 3%, compounded monthly. How much are you required to pay at the end of each month, starting with the end of the first month?

Exercise 4.1.24. You take out a mortgage of $225 000 over 25 years at a nominal annual interest rate of 3%, compounded monthly. How much are you required to pay at the end of each month, starting with the end of the first month?

Exercise 4.1.25. You plan to take out a mortgage for a new house over 25 years at a nominal annual interest rate of 6%, compounded monthly. You plan to be able to afford to pay $800 at the end of each month. What is the biggest mortgage you can afford?

Exercise 4.1.26. You plan to take out a mortgage for a new house over 30 years at a nominal annual interest rate of 9%, compounded monthly. You plan to be able to afford to pay $550 at the of each month. What is the biggest mortgage you can afford?

Exercise 4.1.27. You turned 25 today and would like invest some money in an annuity that pays you $1000 a month from the day you turn 60 until the day you turn 80 (including that day). The nominal annual interest rate is 6%, compounded monthly. How much do you need to put into the account today?

Exercise 4.1.28. You turned 25 today and would like invest some money in an annuity that pays you $1000 a month from the day you turn 70 until the day you turn 90 (including that day). The nominal annual interest rate is 6%, compounded monthly. How much do you need to put into the account today?

Exercise 4.1.29. How much interest do you end up paying on the mortgage in Exercise 4.1.23?

Exercise 4.1.30. How much interest do you end up paying on the mortgage in Exercise 4.1.24?

Exercise 4.1.31. Find the sum of the series

$$2 + 6 + 18 + 54 + \ldots + 13\,122.$$

Exercise 4.1.32. Find the sum of the series

$$2 - 6 + 18 - 54 + \ldots + 13\,122.$$

Exercise 4.1.33. Find the sum of the series

$$-2 + 6 - 18 + 54 - \ldots - 13\,122.$$

Exercise 4.1.34. Find the sum of the series

$$1 + \frac{1}{2} + \frac{1}{4} + \frac{1}{8} + \ldots + \frac{1}{128}.$$

Exercise 4.1.35. Find the sum of the series

$$1 - \frac{1}{2} + \frac{1}{4} - \frac{1}{8} + \ldots - \frac{1}{128}.$$

Exercise 4.1.36. Find the sum of the infinite series

$$1 + \frac{1}{2} + \frac{1}{4} + \frac{1}{8} + \ldots$$

Exercise 4.1.37. Find the sum of the infinite series

$$1 - \frac{1}{2} + \frac{1}{4} - \frac{1}{8} + \ldots$$

Exercise 4.1.38. Longevity is in both your maternal and paternal blood lines, so you decide to purchase a perpetuity that pays you \$12 000 at the end of every year. The interest rate for the perpetuity is 7% per year. How much do you need to put into the perpetuity account?

Exercise 4.1.39. You decide to purchase a perpetuity that pays you \$800 at the end of every month. The interest rate for the perpetuity is at a nominal annual interest rate of 6% per year. How much do you need to put into the perpetuity account?

Exercise 4.1.40. What happens to an infinite geometric series

$$a + ar + ar^2 + \cdots + ar^n + \cdots$$

when $r > 1$? Does it ever add up to a number? When?

Exercise 4.1.41. What happens to an infinite geometric series

$$a + ar + ar^2 + \cdots + ar^n + \cdots$$

when $r < -1$? Does it ever add up to a number? When?

Inflation is the rate at which the cost of basic items goes up over a year. If the annual inflation rate is 5%, then over a year, an item that cost \$100 at the beginning of the year would cost 5% more, that is, $\$100 + 0.05 \times \$100 = \$105$. Answer Exercises 4.1.42 to 4.1.45.

Exercise 4.1.42. If the annual inflation rate is 6%, what would an item that costs \$250 at the beginning of the year cost at the end of the year?

Exercise 4.1.43. If the annual inflation rate is 9%, what would an item that costs \$4.99 at the beginning of the year cost at the end of the year?

Exercise 4.1.44. Suppose the annual inflation rate is 6%. You invest \$1000 in an interest account that pays at an effective annual rate of 7%. A certain consumer item that costs \$1000 at the beginning of the year ends up costing \$1060 at the end of the year.
(a) How much do you earn over the year?
(b) The investment of \$1000 was enough to buy the item at the start of the year. Was the investment enough to buy the consumer item at the end of the year? More than enough? Less than enough?
(c) Did your investment "beat" inflation?

Exercise 4.1.45. Suppose the annual inflation rate is 8%. You invest \$2000 in an interest account that pays at an effective annual rate of 5%. A certain consumer item that costs \$2000 at the beginning of the year ends up costing \$2160 at the end of the year.
(a) How much do you earn over the year?
(b) The investment of \$2000 was enough to buy the item at the start of the year. Was the investment enough to buy the consumer item at the end of the year? More than enough? Less than enough?
(c) Did your investment "beat" inflation?

4.2 Risk Taking and Probability

Money is only part of what we take into account for our decisions. Risk also plays a huge role. We are ready to take certain chances —we get up and head off to classes and to do chores, even though there is a small possibility of various bad incidents happening to us, ranging from stubbing our toes to car accidents and items falling from the sky and hitting us! Even getting more money involves some risk. You can invest in safe government bonds that pay a fixed but small rate of interest or be riskier and invest in the stock market where you might earn a lot of money or lose a lot. Nothing is completely safe, but we have some idea that randomness plays a role in our lives and we need to take that into account.

So let's turn our attention to probabilities. The **probability** of an event is the chance that it will happen. For some events, that are well defined and for which we know a lot about the possible outcomes and their chances of occurring, the probability can be calculated easily. For example, when we toss a (fair) coin or roll a fair die, we know that each possible outcome is equally likely to appear—a head is as likely as a tail to be shown, and each of $1, 2, 3, 4, 5, 6$ are just as likely to come up on top. If the total probability of something happening is 1, then we see that if the probability of a head turning up in a flip of a coin is x, then the probability of a tail turning up is also x. Since one of the two—heads or tails—must happen and they can't happen at the same time, we find that $x + x = 1$, that is $2x = 1$, so $x = 1/2$, and of course you already know that if you flip a coin, each of heads and tails comes up roughly half the time, that is, the probability of each is $1/2$.

I say "(fair)" before coin to indicate that it is not weighted in any way that would make one side more likely to come up than the other (magicians and nefarious people can have access to unfair coins, or even two sided coins with the head or tail on both sides!). The default when talking about usual objects like coins, dice, decks of playing cards, etc., is that they are "fair," with every outcome just as likely to occur.

When flipping a coin, there are two equally likely outcomes, and hence each outcome has the same probability, namely $1/2$. In general:

> If you have n equally likely outcomes, the probability of any particular one of them occurring is $1/n$.

Thus, for example, if you roll a die, a "three" comes up with probability 1/6 (as do each of the other five numbers). If you pick a card at random from a well shuffled deck, the probability of getting any specific card, say the queen of hearts, is 1/52. When all outcomes are equally likely, the probability of something occurring is simply the number of relevant outcomes divided by the total number of outcomes. So, for example, the probability of drawing a heart from a deck is $13/52 = 1/4$, as there are 13 hearts in the deck, and each of the 52 cards is equally likely to be drawn.

Making It Count

In a variety of settings all outcomes are indeed likely to occur, so in order to calculate probabilities the question centers on counting issues. Sometimes it is possible to count all objects by listing them, but often there are so many to list that we only want to know how many there are, and not list them. So we'll set aside some time right now to talk about how to count.

Let's start off with a simple example. If we toss a coin and then roll a die, how many possible outcomes are there? We can think of this in two stages—we toss the coin, which has two possible outcomes, heads or tails, and then roll the die. For each of the two outcomes of the coin, there are six outcomes for the die, so we get $6 + 6 = 2 \times 6 = 12$ many outcomes— we *multiply* the number of outcomes for each when we can do things sequentially. This leads to our first counting principle:

The Multiplicative Principle of Counting: Suppose you can break up your counting into a sequence of events, where you have a ways to make your first choice, then, having done that, b subsequent choices for the second event, then having done those, c subsequent choices, and so on. Then the total number of items is the *product* of all the different choices:

$$a \times b \times c \cdots$$

On the other hand, if we either toss a coin *or* roll a die (but not both!), then we have two cases: either we choose the coin and toss it, with two possible outcomes, <u>or</u> we choose the die and roll it, with six possible outcomes, and we see that from our cases, we have $2 + 6 = 8$ possible outcomes—the *sum* of the number of outcomes for each case. Here we have an illustration of our second counting principle:

The Additive Principle of Counting: Suppose you can break up your counting into a bunch of non-overlapping cases, where you can count a items in the first case, b items in the second case, c items in the third case, and so on. Then the total number of items is the sum of all the different items in each of the cases:

$$a + b + c + \cdots$$

Let's practice using the principles by solving a few counting problems. The numbers may surprise you! Let's begin by counting how many license plates you can make of three letters followed by three numbers.

We can think of listing these by first choosing the first letter in the license plate, then the second letter, then the third, then the first number, the second number and finally the third number.

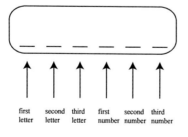

The sequential nature of the counting suggests we use the

multiplicative principle. There are 26 choices for the first letter. Having chosen the first letter, there are still 26 choices for the second letter, and then 26 for the third letter again. After having chosen all three letters, there are 10 choices for the first number (0 through 9), then 10 choices for the second number, and finally 10 for the third number.

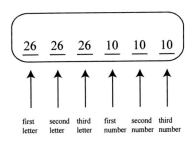

The multiplicative principle says that the total number of ways we can make our choices (which is the number of license plates) is the product of these numbers

$$26 \times 26 \times 26 \times 10 \times 10 \times 10 = 17,576,000.$$

Depending in the state you live in, this may not be enough license plates, and if so, your state has already moved to increase the number of letters or numbers in your license plates (you get a bigger bang for the buck if you add another letter rather than a number, as there are 26 choices for a letter while only 10 for a number).

Now suppose your state has decided to issue some special plates as well to select government workers, starting those plates with the two letters "GV" followed by four nonzero digits (that is, each digit must be 1, 2, 3, 4, 5, 6, 7, 8 or 9). How many plates are available now?

This is a perfect opportunity to apply the additive principle, as there are essentially two cases:

- the regular license plates (consisting of three letters followed by three digits), and

- government plates (consisting of "GV" followed by four digits).

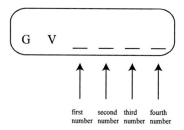

first second third fourth
number number number number

We have seen that there are 17 576 000 of the former. Counting the latter is another instance of the multiplicative principle: for each of the first digits following "GV" there are nine choices (any one of the numbers 1 through 9), then, having made a choice, nine choices for the next digit, and so on. This yields $9 \times 9 \times 9 \times 9 = 6561$ new plates for government employees. By the additive principle, the total number of license plates is

$$17\,576\,000 + 6561 = 17\,582\,561.$$

So we see that both the multiplicative and additive principles can be used in counting problems, and you'd be surprised at just how often just these two rules can be used.

Sometimes we can think of our counting as ordering the objects sequentially—one after another, in order—while at other times we merely select objects, with the order unimportant. This takes us to two terms you probably heard back in high school—permutations and combinations.

Suppose there is a set of n different items. A *permutation* of k of the items is an *ordering* of k different items from the set. A *combination* of k of the items is a selection of k different items from the set (the order of the items is unimportant). The number of permutations of k items from n items is denoted by $P_{n,k}$, while the number of combinations of k items from n items is denoted by $C_{n,k}$ (or $\binom{n}{k}$).

Counting permutations is a straight application of the multiplicative principle. For to order k different items from the n, we choose first the first item, in n ways, as there are n different items. Having chosen this first item, there are now $n - 1$ ways to choose the second item (as you can't use the first item chosen). We proceed to successively choose a new item, different from all of the previous ones. The formula for the number of choices for the last (kth) position might seem a little odd, but remember, by the time we try to fill that position, we have used up $k - 1$ items (and hence they are the unavailable ones at this point). The multiplicative principle implies that the number of such orderings, that is, the number of permutations of k items from the n items, is the product of these numbers, so we have

$$P_{n,k} = n \times (n - 1) \times (n - 2) \times \cdots \times (n - (k - 1)).$$

So, for example, the number of ways to play out 5 cards, from left to right in order, from a deck of 52 is a permutation of $k = 5$ items from $n = 52$ objects, and hence is the product of 52 times 51 and so on all the way down to $52 - (5 - 1) = 48$:

$$P_{52,5} = 52 \times 51 \times 50 \times 49 \times 48 = 311\,875\,200.$$

We notice that in the formula for $P_{n,k}$, the list of numbers in the product starts at n and works its way down. We let the factorial of n, written as $n!$, denote the product of all the positive integers up to and including n:

$$n! = n \times (n - 1) \times (n - 2) \times \cdots \times 2 \times 1.$$

Then we can rewrite $P_{n,k}$ by multiplying the formula by the rest of the numbers down to 1 and dividing by the same new numbers to cancel them out (it is not as useless as it sounds!):

$$
\begin{aligned}
P_{n,k} &= n \times (n - 1) \times (n - 2) \times \cdots \times (n - (k - 1)) \\
&= \frac{n \times (n - 1) \times (n - 2) \times \cdots 2 \times 1}{(n - k) \times (n - k - 1) \times \cdots 2 \times 1} \\
&= \frac{n!}{(n - k)!}
\end{aligned}
$$

So we have a new concise formula for the number of permutations of k items from n:

$$P_{n,k} = \frac{n!}{(n-k)!}.$$

Back to our example of ordering 5 cards out of a deck of 52, we find we can do this in

$$P_{52,5} = \frac{52!}{(52-5)!} = \frac{52!}{47!}$$

many ways. To evaluate this, we realize that $52! = 52 \times 51 \times 50 \times 49 \times 48 \times 47!$, so

$$
\begin{aligned}
P_{52,5} &= \frac{52 \times 51 \times 50 \times 49 \times 48 \times \cancel{47!}}{\cancel{47!}} \\
&= 52 \times 51 \times 50 \times 49 \times 48 \\
&= 311\,875\,200,
\end{aligned}
$$

as before. What you <u>don't</u> want to do is to work out the factorials on the top and bottom first and then try to cancel, as factorials get large very quickly and will exceed the capability of any calculator you use. Use your math!

Armed with this formula and the multiplicative principle, we can make short work out of a formula for combinations. For to order k items from n, we can first select k of them (in $C_{n,k}$ ways) and then order them. But the number of ways you can order k different items is $P_{k,k} = k!$ The multiplicative principle says that the number of ways we select first then order them is the product $C_{n,k} \times k!$, but this is the same as $P_{n,k} = \frac{n!}{(n-k)!}$, so we have that $C_{n,k} \times k! = \frac{n!}{(n-k)!}$, or

$$C_{n,k} = \frac{n!}{(n-k)!k!}.$$

As an example, if we want to count the number of ways to draw a *hand* of 5 cards from a deck of 52 (a standard poker

For this formula to work, we need to make sense out of $0!$, which we define mysteriously to be 1. Actually, it isn't so mysterious, as we know that $P_{n,n}$ is the number of ways to arrange all n items from the n, and is $n \times (n-1) \times (n-2) \times \cdots \times 2 \times 1$, which is $n!$, and so if we want the formula above to hold for $P_{n,n}$, we need $\frac{n!}{(n-n)!} = \frac{n!}{0!}$ to be the same as $n!$, and this means that $0!$ has to be defined as 1.

hand), this is a selection of 5 items from 52 different objects, where the order is unimportant (we don't care what order the 5 cards are given to us). So the number of such hands is

$$
\begin{aligned}
C(52,5) &= \frac{52!}{(52-5)!5!} \\
&= \frac{52!}{47!5!} \\
&= \frac{52 \times 51 \times 50 \times 49 \times 48 \times \cancel{47!}}{\cancel{47!}5 \times 4 \times 3 \times 2 \times 1} \\
&= \frac{311\,875\,200}{120} \\
&= 2\,598\,960,
\end{aligned}
$$

a big number, not nearly as big as $311\,875\,200$, the number of ways to select 5 cards from 52, when the order *is* important. We remark that once again, it is crucial when working out combinations and permutations to cancel early, as the numbers can be enormous otherwise. You must be able to cancel to get a positive integer—even though the formulas for combinations and permutations involve fractions, we know that they count something and must therefore be whole numbers in the end!

Armed with permutations and combinations, we can find a lot of interesting probabilities. For example, what is the probability that in poker we draw a flush, that is, five cards, all of the same suit? We have just seen that there are $2\,598\,960$ equally likely five card hands, so to calculate the probability, we simply need to know how many of these are flushes. To get a flush, we can think of doing two things in order—pick the suit, and then pick the five cards from that suit. We can pick the suit in four ways (there are four suits—hearts, diamonds, clubs and spades), and then we can choose the 5 cards from the 13 of that suit. The latter can be done in $C_{13,5}$ ways. So by the multiplicative principle, the number of flushes is

$$
\begin{aligned}
4 \times C_{13,5} &= 4 \times \frac{13!}{(13-5)!5!} \\
&= 4 \times \frac{13!}{8!5!} \\
&= 4 \times \frac{13 \times 12 \times 11 \times 10 \times 9 \times \cancel{8!}}{\cancel{8!}5!}
\end{aligned}
$$

$$= 4 \times \frac{154\,440}{120}$$
$$= 5148.$$

As all $2\,598\,960$ hands are equally likely, the probability of drawing a flush is

$$\frac{5148}{2\,598\,960} = \frac{33}{16\,660} = 0.00198,$$

pretty unlikely (it happens less than 1% of the time).

Back to the Main Event

In general we talk about an **event** occurring, which is some collection of outcomes. For example, when rolling a die, we might ask about the event of the number showing on top being odd—this corresponds to the outcomes 1, 3 or 5. Events have probabilities of occurring. For example, the probability that the number shown on a roll of a die is odd is $3/6 = 1/2$, as three of the equally likely six outcomes are odd. We write $\mathrm{Prob}(E)$ for the probability that event E occurs, so $\mathrm{Prob}(\text{number on top is odd}) = 1/2$.

Events that are sure to occur (such as a roll of a die showing a number less than 10 on top or the sun coming up tomorrow) have probability 1; events that never happen (such as rolling a 7 on a die or seeing a green unicorn) have probability 0. The probability of any event is always between 0 and 1 (though it can be 0 or 1). For an event E, we let $\overline{E}$ denote the opposite or **complementary event**, that is, that E doesn't occur; for example, if E is the event that a rolled die shows an odd number, then $\overline{E}$ denotes the event that a rolled die *doesn't* show an odd number on top (that is, that it shows an even number). A bit more notation: for events E and F, $E \cup F$ is the event that E *or* F (or both) occurs, while $E \cap F$ is the event that *both* E and F occur.

There are a few rules about calculating the probabilities of events that are worth knowing. Here is the first one.

The symbols $\cup$ and $\cap$ are the symbols for *union* and intersection of sets, as we can view events as really being sets, that is, collections, of objects, namely, basic outcomes.

Suppose that E and F represent two events. Then

$$\text{Prob}(E \cup F) = \text{Prob}(E) + \text{Prob}(F) - \text{Prob}(E \cap F)$$

so if E and F are **mutually exclusive**, that is, they both can't happen at the same time, then

$$\text{Prob}(E \cup F) = \text{Prob}(E) + \text{Prob}(F).$$

So, for example, when drawing a card at random from a standard deck of 52 cards, consider the probability of drawing either a club or an ace.

$$
\begin{aligned}
\text{Prob(club or ace)} &= \text{Prob(club)} + \text{Prob(ace)} - \text{Prob(club and ace)} \\
&= 13/52 + 4/52 - 1/52 = 16/52 = 4/13.
\end{aligned}
$$

On the other hand, the probability of drawing either an ace or a jack is

$$
\begin{aligned}
\text{Prob(ace or jack)} &= \text{Prob(ace)} + \text{Prob(jack)} - \text{Prob(ace and jack)} \\
&= 4/52 + 4/52 = 8/52 = 2/13,
\end{aligned}
$$

since the events of drawing an ace and of drawing a jack are mutually exclusive (they can't happen at the *same* time). Thus drawing an ace or a jack is half as likely as drawing a club or an ace.

A more complicated example is the following: what is the probability of drawing a *flush* (all of the same suit) or a *straight* (all cards in sequential order, with an ace allowed to be either at the top or bottom)? Let F denote the event of getting a flush, and S the event of getting a straight. We want to find $\text{Prob}(F \cup S)$. We have seen previously that the probability of drawing a flush is

$$\text{Prob}(F) = 5148/2\,598\,960$$

(we try not to approximate with decimals until the end, to avoid rounding errors). We can work out the probability of a straight by counting how many straights there are. And to do so, what we do is first pick the number of lowest card of the straight, in 10 ways (from ace, 2, 3, 4, 5, 6, 7, 8, 9, 10)—once we do this, the numbers of the straight are completely determined (as being the 5 consecutive numbers starting at the lowest one). But then we need to pick a suit for each of these numbers, and we can do so sequentially in $4 \times 4 \times 4 \times 4 \times 4 = 1024$, by the multiplicative principle. This principle also tells us that the number of straights is $10 \times 1024 = 10\,240$ so the probability of a straight is

$$\text{Prob}(S) = \frac{10\,240}{2\,598\,960}.$$

We also need to find $\text{Prob}(F \cap S)$, the probability of getting both a flush and a straight, at the same time (this is called drawing a *straight flush*). How many such hands are there? To count these, we pick first the suit (in 4 ways), and then pick the lowest card of the straight (in 10 ways, as before). So there are $4 \times 10 = 40$ straight flushes, and

$$\text{Prob}(F \cap S) = \frac{40}{2\,598\,960}.$$

We put these all together:

$$
\begin{aligned}
\text{Prob(a flush or a straight)} &= \text{Prob}(F \cup S) \\
&= \text{Prob}(F) + \text{Prob}(S) - \text{Prob}(F \cap S) \\
&= \frac{5148}{2\,598\,960} + \frac{10\,240}{2\,598\,960} - \frac{40}{2\,598\,960} \\
&= \frac{15\,348}{2\,598\,960} \\
&= \frac{1279}{216\,580} \\
&= 0.006.
\end{aligned}
$$

Thus the chances of getting a flush or a straight in a poker hand is 0.006—it happens only about 0.6% of the time.

Here is a very important notion, one that we return to again and again. We say that events E and F are **independent events** if whether E occurs or not in no way affects whether F occurs. For example, if you toss a coin and roll a die, and E is the event that the coin comes up heads and F is the event that the die shows a 5, then as the outcome of the toss of the coin and the roll of the die in no way affect one another, events E and F are independent.

Independence is a very powerful tool to use when calculating probabilities. The fundamental rule for the probability of independent events is the following:

Suppose that E and F represent two *independent* events. Then

$$\text{Prob}(E \cap F) = \text{Prob}(E) \times \text{Prob}(F).$$

So for tossing a coin and rolling a die,

$$\text{Prob}(\text{coin shows heads and die shows 5})$$
$$= \text{Prob}(\text{coin shows heads}) \times \text{Prob}(\text{die shows 5})$$
$$= \frac{1}{2} \times \frac{1}{6}$$
$$= \frac{1}{12}.$$

Similarly, if you roll two dice, then the probability that you get "snake eyes" (that is two 1's) is $1/6 \times 1/6 = 1/36$, as the outcomes of the rolls of the first and second dice are independent. On the other hand, if you ask for the probability of drawing two cards in succession, one after the other, from a deck of 52 and ask for the probability that both are diamonds, then even though the probability that you draw a diamond on a single draw of a card is $13/52 = 1/4$, the probability that both are diamonds is <u>not</u> $1/4 \times 1/4 = 1/16$, as the events of drawing a diamond on the first card and drawing on the second card are <u>not</u> independent. They are **dependent events**—the chance of drawing a diamond on the second card goes down slightly if you've drawn a diamond on the first card (and up slightly if you don't), compared to drawing a diamond on the second card if you didn't draw a diamond on the first. Had we noted the first card and replaced it in the deck and shuffled several times, indeed the two events would be independent, but such is not the case.

Of course, we can calculate the probability directly that both cards are diamonds, as there are $52 \times 51 = 2652$ equally likely sequences of two cards from the deck (this is a permutation of 2 items taken from 52). Of these, $13 \times 12 = 156$ are both diamonds, again by the multiplicative principle. So the probability that both are diamonds is $156/2652 = 1/17$, which is a bit smaller than the $1/16$ had the events of a diamond on the first card and a diamond second card been independent.

But there is another way to calculate this, based on a variant of standard probabilities of events. It is called **conditional probability**, where we ask the probability of an event, *given* that another has occurred. We write $F|E$ to denote the event of F *given* that E has occurred and call $\text{Prob}(F|E)$ the probability of F, given E. So if someone drew a card from a standard deck and told you that the card was a red card, what would be the probability that the card is the jack of diamonds? Certainly, if you had no information, the probability that the card was the jack of diamonds would be $1/52$, but it becomes more likely if you know that the card is red. Knowing that the card is red reduces the choices to 26 equally likely cards, and only one of these is the jack of diamonds, so

$$\text{Prob}(\text{card is the jack of diamonds}|\text{card is red}) = \frac{1}{26}.$$

But note that

$$\frac{1}{26} = \frac{1/52}{1/2}$$

so that

$$\begin{aligned} &\text{Prob}(\text{card is the jack of diamonds}|\text{card is red}) \\ &= \frac{\text{Prob}(\text{card is the jack of diamonds})}{\text{Prob}(\text{card is red})}. \end{aligned}$$

This isn't just an accident. Here is the way we can calculate conditional probabilities:

Suppose that E and F represent two events, with $\text{Prob}(F) \neq 0$. Then

$$\text{Prob}(E|F) = \frac{\text{Prob}(E \cap F)}{\text{Prob}(F)}.$$

It follows that, provided $\text{Prob}(F) \neq 0$, events E and F are independent just in the case $\text{Prob}(E|F) = \text{Prob}(E)$, that is, the information that F has occurred in no way affects the probability that E has occurred, since if E and F are independent, then

$$
\begin{aligned}
\text{Prob}(E|F) &= \frac{\text{Prob}(E \cap F)}{\text{Prob}(F)} \\
&= \frac{\text{Prob}(E) \times \text{Prob}(F)}{\text{Prob}(F)} \\
&= \text{Prob}(E),
\end{aligned}
$$

while if $\text{Prob}(E|F) = \text{Prob}(E)$, then

$$
\text{Prob}(E) = \text{Prob}(E|F) = \frac{\text{Prob}(E \cap F)}{\text{Prob}(F)}
$$

and by cross-multiplying,

$$
\text{Prob}(E \cap F) = \text{Prob}(E) \times \text{Prob}(F).
$$

Returning now to our problem of calculating the probability that when two cards are sequentially drawn from a deck both are diamonds, we let F be the event that the first card drawn is a diamond and E be the event that the second card is a diamond then what we want is $\text{Prob}(E \cap F)$, which we can rewrite from

$$
\text{Prob}(E|F) = \frac{\text{Prob}(E \cap F)}{\text{Prob}(F)}
$$

as

$$
\text{Prob}(E \cap F) = \text{Prob}(F) \times \text{Prob}(E|F).
$$

Clearly $\text{Prob}(F) = 13/52 = 1/4$. What about $\text{Prob}(E|F)$, that is, what is the probability that the second card is a diamond, given that the first was a diamond? Clearly, if the first card is known to be a diamond, that leaves 51 other equally likely cards left from which to choose 1 of the 12 diamonds left, and hence $\text{Prob}(E|F) = 12/51 = 4/17$. Thus

$$
\begin{aligned}
\text{Prob(both cards are diamonds)} &= \text{Prob}(E \cap F) \\
&= \text{Prob}(F) \times \text{Prob}(E|F) \\
&= \frac{1}{4} \times \frac{4}{17} \\
&= \frac{1}{17},
\end{aligned}
$$

as we found before. But it never, ever hurts in mathematics to have more than one way to do something!

Let's do one more interesting example, one that puts a lot of our good work on probabilities (and more!) to real work. The National Basketball Association (NBA) holds its annual Draft Lottery, where teams select new players from a pool of eligible newcomers. Unlike Major League Baseball (MLB), where the draft order is set by the previous year's standings (MLB teams chose in reverse order of standing), in the NBA randomness is introduced to make the draft more exciting. There are 30 teams in the NBA, with 16 making the playoffs. The draft lottery is as follows: 14 balls, numbered from 1 to 14, are spun around in a machine. The combinations of four numbers from $\{1, 2, \ldots, 14\}$, except for the combination $\{11, 12, 13, 14\}$, are divided up among the 14 teams that don't make the playoffs, with the number of combinations assigned to each team depending on their ranking the previous season. Note that our work on counting tells us that there are $C_{14,4} = 1001$ combinations of 4 numbers from 14, so there are 1000 combinations (a nice number!) to apportion to the teams in question. Table 4.2 shows how many tickets are set aside for each team. The lower place teams have more of a chance to get an early draft pick, but no guarantee. Even the bottom place team has only a 25% chance of getting the first draft pick (though it is better than any other team).

A combination of four numbers from $\{1, 2, \ldots, 14\}$ is a collection of four different numbers from the set, such as $\{1, 4, 7, 13\}$ and $\{2, 5, 9, 14\}$; the set brackets $\{$ and $\}$ indicate that the order of the numbers inside makes no difference.

team's rank	# lottery combinations	probability of first draft pick
30th	250	0.250
29th	199	0.199
28th	156	0.156
27th	119	0.119
26th	88	0.088
25th	63	0.063
24th	43	0.043
23th	28	0.028
22th	17	0.017
21th	11	0.011
20th	8	0.008
19th	7	0.007
18th	6	0.006
17th	5	0.005

Table 4.2: NBA Draft Lottery probabilities. The rules are modified if there are ties in rank—if there are two teams with a tied position at the end of the previous season, then they split the total number of combinations allotted for their ranks, with extras being allotted randomly between them.

Four balls are drawn randomly from the 14 (with a redraw

of all 4 in the unlikely event that $11, 12, 13, 14$ are drawn). The owner of the 4 ball combination gets the first draft pick and the balls are put back into the machine. Two more draws are done for the second and third draft positions, with a redraw any time a previous winner in the year's lottery combination comes up again (or $11, 12, 13, 14$ comes up). So here is an interesting question—what is the probability that a certain team gets the *second* draft pick? For example, what is the probability that the bottom ranked team (the 30th team) gets the second pick?

Let A be this event, and let $B_{29}, B_{28}, \ldots, B_{17}$ be the events that the 29th, 28th, ..., 17th team, respectively, gets the first draft pick. Then event A is split up into the events

$$B_{29} \cap A, B_{28} \cap A, \ldots, B_{17} \cap A,$$

that is,

$$A = (B_{29} \cap A) \cup (B_{28} \cap A) \cup \ldots \cup (B_{17} \cap A),$$

as if the 30th team gets the second pick, then one of the other teams, ranked 29th to 17th, must get the first pick and <u>then</u> the 30th team must get the next pick. We can safely ignore the draw of the $11, 12, 13, 14$ combination, as it only stalls the process and has no effect otherwise.

As the events $B_{29} \cap A, B_{28} \cap A, \ldots, B_{17} \cap A$ are pairwise disjoint (no two can happen at the same time), it follows that

$$\text{Prob}(A) = \text{Prob}(B_{29} \cap A) + \text{Prob}(B_{28} \cap A) + \cdots + \text{Prob}(B_{17} \cap A).$$

This turns our one probability into a sum of 13 others, but we'll see that the latter are easier to calculate. Let's look at the first term, $\text{Prob}(B_{29} \cap A)$, the probability that the 29th team gets the first pick and the 30th team gets the second pick. We can rewrite this with conditional probabilities:

$$\text{Prob}(B_{29} \cap A) = \text{Prob}(B_{29}) \times \text{Prob}(A | B_{29}),$$

with the right side being the probability of the 29th team getting the first pick (which from the table is 0.199), and the 30th team getting the next pick, given that the 29th team got the first pick. This amounts to the event that the only combinations that come up after the first one are those for the 29th team, until one for the 30th team is drawn. So what about this last event? We can break it up into infinitely many pairwise disjoint events, namely,

- one of the 30th team's combinations is drawn after the first draw, or

- the second draw is one of the 29th team's, and then on the next draw one of the 30th team's combinations is drawn after the first draw, or

- the second and third draws are both one of the 29th team's, and then on the next draw one of the 30th team's combinations is drawn after the first draw,

- the second, third and fourth draws are all one of the 29th team's, and then on the next draw one of the 30th team's combinations is drawn after the first draw,

and so on. These probabilities, though there are infinitely many of them, are easy to calculate:

- the probability that one of the 30th team's combinations is drawn after the first draw is 0.250,

- the probability that the second draw is one of the 29th team's, and then on the next draw one of the 30th team's combinations is drawn after the first draw is $0.199 \cdot 0.250$, as the draws are independent,

- the probability that the second and third draws are both one of the 29th team's, and then on the next draw one of the 30th team's combinations is drawn after the first draw is $0.199 \cdot 0.199 \cdot 0.250$,

- the probability that the second, third and fourth draws are all one of the 29th team's, and then on the next draw one of the 30th team's combinations is drawn after the first draw is $0.199 \cdot 0.199 \cdot 0.199 \cdot 0.250$,

and so on. It follows that the probability that the 29th team gets the first pick and the 30th team gets the second is

$$
\begin{aligned}
\text{Prob}(B_{29} \cap A) &= \text{Prob}(B_{29}) \times \text{Prob}(A|B_{29}) \\
&= 0.199 \left(0.250 + 0.199 \cdot 0.250 + 0.199^2 \cdot 0.250 + \right. \\
&\qquad \left. 0.199^3 \cdot 0.250 + \cdots \right)
\end{aligned}
$$

The right side looks impossible to add up, with infinitely many terms, but wait—in the previous section on money, we saw exactly how to add up such infinite series, if they had a common ratio between -1 and 1. The series in the parentheses is indeed a geometric one, with the first term $a = 0.250$ and the common ratio $r = 0.199$, so it sums to

In our consideration of perpetuities, we learned to add up such infinite series.

$$
\frac{a}{1-r} = \frac{0.250}{1-0.199}.
$$

Thus we find that

$$
\begin{aligned}
\text{Prob}(B_{29} \cap A) &= 0.199 \left(0.250 + 0.199 \cdot 0.250 + 0.199^2 \cdot 0.250 + \right. \\
&\qquad \left. 0.199^3 \cdot 0.250 + \cdots \right) \\
&= 0.199 \cdot \frac{0.250}{1-0.199} \\
&= 0.250 \cdot \frac{0.199}{1-0.199}.
\end{aligned}
$$

We can find the other probabilities in a similar way. Thus we find that

$$
\begin{aligned}
\text{Prob}(A) &= \text{Prob}(B_{29} \cap A) + \text{Prob}(B_{28} \cap A) + \cdots + \text{Prob}(B_{17} \cap A) \\
&= 0.250 \cdot \frac{0.199}{1-0.199} + 0.250 \cdot \frac{0.156}{1-0.156} + 0.250 \cdot \frac{0.119}{1-0.119} + \\
&\quad 0.250 \cdot \frac{0.088}{1-0.088} + 0.250 \cdot \frac{0.063}{1-0.063} + 0.250 \cdot \frac{0.043}{1-0.043} + \\
&\quad 0.250 \cdot \frac{0.028}{1-0.028} + 0.250 \cdot \frac{0.017}{1-0.017} + 0.250 \cdot \frac{0.011}{1-0.011} + \\
&\quad 0.250 \cdot \frac{0.008}{1-0.008} + 0.250 \cdot \frac{0.007}{1-0.007} + 0.250 \cdot \frac{0.006}{1-0.006} + \\
&\quad 0.250 \cdot \frac{0.005}{1-0.005} \\
&= 0.215.
\end{aligned}
$$

Therefore, the probability that in the NBA draft the 30th team gets the *second* draft pick is 0.215, that is, 21.5% of the time, less than its chance of getting the first draft pick. And the probability of the 30th team getting one of the first two draft picks is $0.250 + 0.215 = 0.465$, or 46.5% of the time.

You can calculate the probabilities of the other teams getting the second draft pick similarly, and move on to third draft picks as well if you like. The point I want to drive to the basket with this is that we can use our understanding of probabilities to calculate chances in the draft for the different teams, and we've reused some of our newfound knowledge about infinite geometric series naturally along the way.

Probabilities in Real Life

Randomness permeates games like poker, blackjack and craps, where it is fairly clear why the process is completely random and how to calculate the probabilities. But in real life things are a bit murkier, and we base our probabilities on the **frequency** of the event, that is, the proportion of those in the population for which the event has occurred. For example, what is the probability that a random person in the U.S. will be killed by lightning during a year? In 2010, there were 29 deaths due to lightning strikes. As the U.S. population that year was $310\,232\,863$, the probability of a random individual during that year being killed by lightning was $29/310\,232\,863$, or about 0.0000000935—it is extraordinarily rare (but does happen).

Would this have been your chance of being killed by lightning that year? That is a tougher question to answer. You aren't a random person. You may be more cautious, refusing to go out in storms, and so your probability would have been even smaller. On the other hand, if you were an avid camper, a bit of a thrill-seeker who liked storm chasing, it might be higher. But in the absence of any other information, we would take $29/310\,232\,863$ as your probability of being killed by lightning that year (as you are reading this textbook, written after 2010, congratulations—you survived!).

Working with such real-life probabilities has been extraordinarily useful, connecting up statistics and probability theory. You can talk about a person's probability of getting a certain disease, winning a lottery, placing at the Olympics, finding out you're adopted, and so on, based on frequencies in a population that includes yourself. The more information you can take into account when restricting the relevant population, the more accurate the probability will be. For instance, if you want to know your chances of getting lung cancer, Prob(lung cancer), you could look at the probability that a random American gets the disease, as a first attempt. But if you can take into account the city you live in (some cities have a higher prevalence of lung cancer due to pollution and other factors), your previous and present smoking status, your gender and age, and so on, and restrict yourself to the frequency of lung cancer within the relevant subset of the population, you would get a conditional probability, say

Prob(lung cancer|live in Chicago, non-smoker, female, age 24),

which would give you a more accurate estimate of your personal probability of getting lung cancer. The trick is, the more information you try to take into account, the more difficult it is to find and count the relevant population. So we settle somewhere in between, taking into account as much information as we can, and create a probability based on the frequency, basing our decisions on this estimate of the true, unknown probability.

We point out that there is a deep connection between probabilities and frequencies. If you repeat, independently, the same random process over and over and over again, and count the number of times that a certain event occurs each time, then the **Law of Large Numbers** in mathematics states that the frequency of the event happening will approach the probability of the event. So, for example, if you flip a coin many, many, many times, the proportion of heads will approach 1/2, the exact probability of heads, over time. So after say 10 tosses, you might find that you got only 2 heads, but over a 100 tosses, you would be surprised to only get 20% heads—you probably would get something like 57 heads (for a proportion of 0.57). And over say 1000 heads, you would expect the proportion to be even closer to 1/2, say 483 heads (for a proportion of 0.483).

The Folly of Repeated Risk Taking

OK, let's examine some real-life use of probabilities in decision making. First of all, what about safety in airline travel?

Between 1990 and 2000 the death rate from commercial airline accidents was about 0.08 per billion miles flown. So if you haven't traveled very much by plane, then your chance of dying on an airplane is very, very small, and a risk worth taking, I think.

But at some point, risks, even small ones, do add up. I remember the day that Steve Irwin, the great naturalist, died, from a freak accident—his heart was pierced by a stingray's barb. Certainly the probability of this happening was very small, and the world was shocked by the unlikely way he died. But should we have been? Let's think about what probability says. Steve Irwin subjected himself to animal-related risks every day, whether from crocodiles, rhinos, elephants, snakes or other creatures. Suppose that on a given day, Steve, with his unique animal skills, had only a 0.0001 chance of dying in an encounter with an animal. That is, with 99.99% certainty, he would survive the day.

Figure 4.3: The great naturalist, Steve Irwin.

But now let's think about the risk of repeating the risk day after day for 40 years. Let D_i be the event of Steve Irwin living through the ith day, i running from 1 to $7305 = 20 \times 365.25$. The probability that Steve survived the first day is $\text{Prob}(D_1) = 0.9999$. The probability Steve Irwin survived the first two days is

$$\text{Prob}(D_1 \cap D_2) = \text{Prob}(D_1) \times \text{Prob}(D_2|D_1)$$
$$= 0.9999 \times 0.9999$$
$$= 0.9999^2,$$

from our formula for conditional probability, as $\text{Prob}(D_2|D_1)$ is simply the probability that Steve Irwin lived through day 2, given he survived day 1. Similarly, the probability that Steve Irwin lived through day 3 is

$$\text{Prob}(D_1 \cap D_2 \cap D_3) = \text{Prob}(D_1 \cap D_2) \times \text{Prob}(D_3|D_1 \cap D_2)$$
$$= 0.9999 \times 0.9999 \times 0.9999$$
$$= 0.9999^3,$$

and so on. Thus the probability of Steve Irwin surviving 20 years, that is, all 7305 days, is $0.9999^{7305} = 0.4816$. That is, he had more than a 50% chance of dying in some animal encounter over the 20 year period! The lesson learned is this:

Life Lesson: Taking an occasional risk is quite safe, but repeated risk taking is an accident waiting to happen.

Moreover, rare events happen to everyone—you can't avoid the numbers. What those rare events, good or bad, will be is impossible to know. But rest assured, some rare event will happen to you. All you can do is hedge your bets and avoid some specific outcomes by staying away from risky behavior.

Be Less Afraid!

Conditional probabilities can be even more useful when we have to decide whether to panic or not over medical test results. I remember one Friday when I found a lump in my neck. I went into a walk-in clinic, and after examining the enlarged node, the doctor said "I'm going to run some tests. It could be lymphoma." I spent the whole weekend worrying about the "C" word.

In retrospect, I should have been less worried than I was, and to explain why, I need to bring up a very powerful rule about conditional probabilities:

Bayes' Theorem: Suppose that E and F represent two events, with $\text{Prob}(F) \neq 0$. Then

$$\text{Prob}(F|E) = \frac{\text{Prob}(E|F)\text{Prob}(F)}{\text{Prob}(E|F)\text{Prob}(F) + \text{Prob}(E|\overline{F})\text{Prob}(\overline{F})}.$$

Bayes' Theorem holds as the top of the fraction on the right is just $\text{Prob}(E \cap F)$, while the bottom is $\text{Prob}(E \cap F) + \text{Prob}(E \cap \overline{F})$, that is, the probability that E happens and F happens plus the probability that E happens and F doesn't happen, which is obviously just the probability that E happens.

So armed with Bayes' theorem, let's go back to my node. We'll let N denote the event that I have a prominent node in my neck and L the event that I have lymphoma. The incidence of lymphoma for my age group at the time was about 2.9 cases

Figure 4.4: Thomas Bayes (1701–1761) was a statistician and a Presbyterian minister.

per 100 000, so we'll take $\text{Prob}(L) = 2.9/100\,000 = 0.000029$, and hence the probability that I didn't have lymphoma was $1 - 0.000029 = 0.999971$ (which makes sense, as having lymphoma is rare).

But what was important to me was not the probability of having lymphoma, but the probability of having lymphoma *given I had an enlarged node*—this was undoubtedly a higher number. So how could I estimate $\text{Prob}(L|N)$? Here is where Bayes' theorem really helps. It states that

$$\text{Prob}(L|N) = \frac{\text{Prob}(N|L)\text{Prob}(L)}{\text{Prob}(N|L)\text{Prob}(L) + \text{Prob}(N|\overline{L})\text{Prob}(\overline{L})}.$$

It looks like we have made the problem worse, trading in one unknown probability for four of them, but the key is whether I have traded one hard one for four much easier ones. We already know that $\text{Prob}(L) = 0.000029$ and $\text{Prob}(\overline{L}) = 0.999971$. What about $\text{Prob}(N|L)$, the probability I have an enlarged node in my neck, given that I have lymphoma? This I would estimate would be very high, perhaps 0.95. And what about $\text{Prob}(N|\overline{L})$, the probability I have an enlarged node in my neck, given that I *don't* have lymphoma? This is smaller, but not too small, as there are many reasons I could have an enlarged node in my neck—an infection, for instance—so I estimate this at about 0.05. So now my estimate for the chances that I have lymphoma, given my node, would be

$$
\begin{aligned}
\text{Prob}(L|N) &= \frac{\text{Prob}(N|L)\text{Prob}(L)}{\text{Prob}(N|L)\text{Prob}(L) + \text{Prob}(N|\overline{L})\text{Prob}(\overline{L})} \\
&= \frac{0.95 \times 0.000029}{0.95 \times 0.000029 + 0.05 \times 0.999971} \\
&= 0.0000029,
\end{aligned}
$$

which is 0.00029%, a minuscule probability. So I didn't have to worry—indeed, lymphoma is very rare, even among those with nodes.

The process I just carried out with Bayes' theorem can help you mentally deal with positive test results for rare diseases, when there are small but positive chances of false positives. Indeed, in many of these cases, positive results are indeed false positives, and you need to keep that in mind before catastrophizing.

Probability and Trickery

Just as we have seen that statistics can be used to confuse, so can probability. One paradox, called the **gambler's fallacy** is a very powerful one, emotionally. It states that after a bad streak, your luck is bound to change soon. If you have lost 10 hands of poker in a row, you feel that you are *due* to win the next hand! This fallacy has led to many the undoing of a seasoned gambler.

The thing is, probability has no memory—the fact that you have lost the previous 10 hands is the same as you having won the previous 10 hands. Neither affects the hands that are to come. Where this deep-felt erroneous belief comes from is our innate appreciation for the Law of Large Numbers, which says that, over time, the frequency of winning has to approach the probability of winning, so indeed you do have to win sometimes (if winning is possible). But the Law is impervious to what has gone on before—we are only talking about limits, when the number of occurrences gets REALLY LARGE, and this may have to wait until the number of new plays of the game dwarfs the number played previously. Keeping the gambler's fallacy in mind is crucial to keep from falling into a trap and losing good money after bad. Having said this, everyone, including mathematicians, still are prone to fall into the trap.

Other traps even get some of the smartest people—lawyers and judges. The law is inherently intertwined with probabilities—having *probable cause for arrest* and finding for one party on the *balance of probabilities* (i.e., when the side's statement of claim is more likely to be true than not true). As legal proof is almost never 100%, sides are left to argue about the likelihood of guilt or innocence. Moreover, if gamblers risk losing their hard-earned cash with the gambler's fallacy, those on trial often have much more at stake.

You may be too young to remember, but there was a famous court case in 1995 about whether football great O.J. Simpson had murdered his ex-wife, Nicole Brown Simpson, and Ronald Goldman, a waiter at a Los Angeles restaurant. (and a famous slow speed police car chase of a white Ford Bronco carrying O.J.). The defense legal team put forward an argument that while there was evidence that O.J. had physically abused his ex-wife while they were married, statistically only a small proportion of women who have been abused in a relationship end up murdered by their partner. Thus

Prob(woman murdered by partner|woman abused by partner)

was small, and hence unlikely. While this probability may indeed be valid, it was irrelevant in the case, as it was known as a fact that the abused ex-wife was indeed murdered. The relevant probability for the case was

Prob(woman murdered by abusive partner|abused woman murdered),

and this is indeed quite high. The wrong conditional probability, though it sounds (in the hands of lawyers) like it is important, was misleading.

Another case shows that problems with conditional probabilities are not the only way even clever people can be misled. In 1998, Sally Clark, a British lawyer, was charged with two counts of murder in connection with the sudden deaths of her two infant sons shortly after their births in the two previous years. While "cot death" (known now as SIDS—Sudden Infant Death Syndrome) was known to be a cause of death of young babies, in court, Professor Sir Roy Meadow, an expert witness for the prosecution, stated that "one sudden infant death in a family is a tragedy, two is suspicious and three is murder unless proven otherwise." He estimated that in a rich, non-smoking household like the Clarks, the probability of a cot death would only be $1/8543$, so that the probability of *two* cot deaths would be, by independence,

$$\text{Prob(two cot deaths in Clark household)} = \frac{1}{8543} \times \frac{1}{8543}$$
$$= \frac{1}{72\,982\,849}$$

or about 1 in 7.3 million, a very rare event (Meadow estimated that two such cot deaths could be expected to occur in England by chance only once in a century). Based on this and on other evidence, Sally Clark was convicted and sentenced to life in prison. The decision held up under appeal.

But statisticians began to raise serious concerns about Meadow's probabilistic arguments. Some suggested that Meadow had cherry-picked (that is selectively chosen) his comparable group— a rich, non-smoking environment—for estimating his frequency probability, but left out the fact that both babies were boys, which raised their likelihood of SIDS (perhaps to 1/300, not 1/8543). Furthermore, was the 1 in 7.3 million the probability that Sally Clark was innocent? Many argued that no, the relevant thing to do is to compare this probability to a competing alternative, of her being committing double infanticide, which is even rarer than two SIDS deaths in the household, and on that basis she should be exonerated.

But perhaps the biggest flaw in Meadow's argument was his belief that the probability of the first baby in the household dying of SIDS and the probability of the second baby in the household dying of SIDS were independent. Does that make sense to you? Not to me. There are many reasons that babies may die of SIDS (some of the reasons are unknown), but there certainly could be a genetic component to such a death, and if so, the probabilities would <u>not</u> be independent—given a first death from SIDS, a second death would much more likely to happen than had the first death from SIDS not occurred.

What happened in the end? A second appeal threw out the conviction, along with much of Professor Meadow's expert evidence. Even though Sally was released in 2003 after 3 years in prison, sadly she died from drinking in 2007, a victim of not only false imprisonment, but bad probabilities as well.

Profiling and Probability

Both in court rooms and outside, people routinely formulate probabilities and accept probabilities that they hear as correct. Suppose we hear or believe, based on news reports, that group A makes up 70% of terrorists, but worldwide, only 0.0001% of group A are terrorists. Seeing a person of group A at the airport, we may feel uncomfortable right away, thinking with probability

0.70 the person is a terrorist. But the probability, while perhaps felt instinctively, is far from correct. We are mistaking the probability that a given terrorist is of group A (which is high, 0.70) with the probability that a person of group A is a terrorist, which is very small (0.00001), and it is the latter that is important at the moment. Not to worry.

But this brings to mind how we formulate probabilities on frequencies, not on real-life data, but data we have gathered subconsciously. Prejudice and racism sometimes have a basis in exactly this type of **statistical inference**. We base our opinions on the opposite gender, on other races, on other religions on experiences. If we have little experience with a group beyond a few individuals, we tend to attribute certain attributes based on the frequency of the attributes among those from the population that we have been exposed to. For example, suppose we have met (or have seen on TV) eight individuals from group B, all acting aggressively. What do we often take as the probability of a person from group B being aggressive? Undoubtedly 1—that is, we feel that 100% of them will act this way. But should we? The frequency of those who act aggressively in the group, from our experience, is eight out of eight, so it looks like we should assign a probability of 1. However, our sample is far from random, and indeed if it is garnered from television news reports, all we see are the extreme cases that make the news—far from a random sample. And without a random sample, the probabilities we form in our minds are subject to bias (and that can be very unfair).

But even if the sample were random, the problem still persists. If we flipped an unknown coin (that is, one for which we have no idea whether it is fair or not) twice and got heads both times, would we think that the probability of getting heads is $2/2 = 1$, that is, we have a two-headed coin? I think not. Actually, there is a result, due to French mathematician Pierre-Simon Laplace (whose name is one of 72 on the Eiffel Tower):

Figure 4.5: Probabilist Pierre-Simon Laplace.

Laplace's Rule: When you independently repeat an experiment with two outcomes, say H or T, n times and find that you get h many H's and t many T's, then, provided that you have no reason to assume any value for the probability of H is more likely than any other, your best estimate for the probability of getting an H is

$$\frac{h+1}{h+t+2}.$$

So in the case where we have flipped a coin twice and ended up with 2 heads, our best estimate for the probability of getting a heads on a toss would be

$$\frac{2+1}{2+0+2} = \frac{3}{4}.$$

In fact, if we tossed the coin 100 times and each time it came up heads, we would estimate the probability that the coin will next come up heads as $(100+1)/(100+0+2) = 101/102$, close to 1, but not 1. And we wouldn't predict that a tail was impossible (with probability 0), but would estimate its probability at $1/102$, small, but not zero.

Returning now to our view of others, even if we found all eight people of group B were aggressive, and even if we felt our sample were indeed a random one, our best estimate for the rate of aggressiveness in group B wouldn't be 1, absolute certainty, but 8/9, which would be close to 1, but not certain. All of our prejudices should be at least tempered by Laplace's Rule.

Finally, the use of statistical inference, from our exposure and experiences, to predict attitudes and characteristics of groups of people is often misleading, but perfectly human. We all do it. So it behooves each of us to act well if we want others not only to think well of us, but of those whose groups we belong to.

Life Lesson: We have the power to positively influence probabilities others form about us, our gender, race, religion and profession, with statistical inference as well.

216

Exercises

For each of Exercises 4.2.1 to 4.2.8 determine whether the outcomes are equally likely.

Exercise 4.2.1. You pick a card from a fully shuffled deck of cards.

Exercise 4.2.2. You pick a card from a new deck of cards.

Exercise 4.2.3. You pick a hand of four cards from a fully shuffled deck of cards.

Exercise 4.2.4. You pick a hand of four cards from a new deck of cards, picking every fourth card from the top.

Exercise 4.2.5. You roll a red die and a white die.

Exercise 4.2.6. You flip a penny, nickel and dime.

Exercise 4.2.7. You pick a person from a classroom by numbering each person then choosing one corresponding to the first number that comes to mind.

Exercise 4.2.8. You pick a person from a classroom by numbering each person then choosing one of the numbers randomly.

Exercise 4.2.9. How many license plates, consisting of three letters followed by three digits, can be made if all the letters must be distinct?

Exercise 4.2.10. How many license plates, consisting of three letters followed by three digits, can be made if all the letters and digits must be distinct?

Exercise 4.2.11. Suppose government plates ("GV" followed by 4 digits) allowed 0 as a digit. What numbers might be misinterpreted as letters and cause confusion on license plates?

Exercise 4.2.12. Your lock has 40 numbered lines on its circular dial. A lock combination consists of 3 numbers from 0 to 39. How many lock combinations are there? How many are there with *distinct* numbers? How secure does that make you feel about your lock?

Exercise 4.2.13. How many ways are there to select a first, second and third card from a deck of 52 cards?

Exercise 4.2.14. How many ways are there to select a hand of 3 cards from a deck of 52 cards?

Exercise 4.2.15. How many ways can we select one president and two vice-presidents from a group of eight people?

Exercise 4.2.16. What is the probability, when you roll a red and white die, that the sum of the pips (the numbers shown on top) is 7?

Exercise 4.2.17. What is the probability, when you roll a red and white die, that the sum of the pips (the numbers shown on top) is 11?

Exercise 4.2.18. What is the probability in poker that you draw four of a kind, that is, four cards, all of the same denomination (for example, four kings)?

Exercise 4.2.19. What is the probability in poker that you draw a full house, that is, two of one denomination, three of another?

Exercise 4.2.20. If you draw a single card from a deck of 52, what is the probability you draw a red card or a face card (i.e., a jack, queen or king)?

Exercise 4.2.21. What is the probability in poker that you draw either a full house or four of a kind?

Exercise 4.2.22. Suppose that $\text{Prob}(E) = 0.3$ and $\text{Prob}(F) = 0.2$.
(a) What is $\text{Prob}(\overline{E})$?
(b) If events E and F are mutually exclusive, what is $\text{Prob}(E \cup F)$?
(c) If events E and F are independent, what is $\text{Prob}(E \cap F)$?
(d) If events E and F are independent, what is $\text{Prob}(E \cup F)$?
(e) If events E and F are independent, what is $\text{Prob}(E|F)$?

Exercise 4.2.23. Which of the following events E and F are independent?
(a) You draw a card and toss a coin. E is the event of getting a flush and F is the event of tossing tails.
(b) You draw two cards, one after the other. E is the event of the first being a king and F is the event of the second being an ace.
(c) You draw one card, look at it, replace it in the deck, shuffle a few times, and pick another card. E is the event of the first being a king and F is the event of the second being an ace.
(d) You select two students at random from a class. E is the event of the first being a girl and F is the event of the second being an girl.
(d) You select one student at random from a class, ask him or her for her list of friends in the class, and then select one of those friends at random. E is the event of the first being a girl and F is the event of the second being a girl.

Exercise 4.2.24. In the NBA draft lottery, what is the probability that the first draft pick goes to one of the bottom four teams, that is, one of the teams ranked 27th to 30th the previous season?

Exercise 4.2.25. In the NBA draft lottery, calculate the probability that the 29th ranked team gets the *second* draft pick.

Exercise 4.2.26. If you flip a coin repeatedly, what is the probability you get heads at some point?

Exercise 4.2.27. If you flip a coin repeatedly, what is the probability you get never get heads?

Exercise 4.2.28. Out of a U.S. population of 318.1 million in 2013, 25.8 million have diabetes. What is the probability that a random person in the U.S. has diabetes?

Exercise 4.2.29. Out of a U.S. population of 318.1 million in 2013, approximately 720 000 have heart attacks. What is the probability that a random person in the U.S. had a heart attack in 2013?

Exercise 4.2.30. Can you multiply your answers to Exercises 4.2.28 and 4.2.29 to find the probability that a random person in the U.S. in 2013 had diabetes *and* a heart attack? Why or why not?

Exercise 4.2.31. Suppose a person's probability of being chosen for a TV talent show is 0.10, and that person has applied and been turned down by 10 shows. Will that person eventually make it onto such a show? Why? Is he more likely to be chosen on the next show he applies to now that he has had such a bad streak of luck?

Exercise 4.2.32. For the person described in Exercise 4.2.31, how many subsequent shows does he have to try out for before his chances get greater than 0.50, that is, more than 50-50?

Exercise 4.2.33. Suppose that a pharmacist has a 0.05 chance of putting the wrong pills in the wrong bottle, a serious problem. The drugstore plans to reduce the problem by having a number of pharmacists check each prescription and will only accept an error probability of at most 0.0001. Complete the following table. How many pharmacists do they need to reduce the error rate below 0.0001?

Number of pharmacists	1	2	3	4	5
Error probability					

Exercise 4.2.34. Suppose that there is a disease for which you are contemplating taking a test. You know that the probability of having the disease is 0.0045. The test comes back positive with probability 0.90 if you have the disease, but with probability 0.15 if you don't. You got a call from the doctor's office that the test was positive. Use Bayes' Theorem to decide your probability of having the disease, given that the test came back positive. Should you worry?

Exercise 4.2.35. Redo Exercise 4.2.34 if the probability of having the disease is 0.05.

For each of the following three problems, describe the error in the use of probability.

Exercise 4.2.36. A lawyer argues that his client couldn't have committed the murder, as his client is under 5 feet tall, and very few murderers are that short.

Exercise 4.2.37. A person gets onto a train that had an accident the day before as it would be very unlikely for a train to have an accident two days in a row.

Exercise 4.2.38. A lawyer argues that the defendant is very likely guilty, as the murder was committed by someone with a limp and wearing a baseball hat, and the defendant had a limp and was wearing such a hat on the day in question.

Exercise 4.2.39. If you have no information about the rate of depression in ex-musicians, and you have met two ex-musicians who were depressed, what is your best estimate for the rate of depression in the group?

And Now the Rest of the Story . . .

The darkened runaway train barreled right into the center of Lac-Mégantic, sparks flying into the night air from the wheels. People stared in disbelief as the tanker cars one by one left the tracks. The oil from the cars ignited in an inferno over a hundred feet high, with multiple explosions as each tanker broke open and the crude caught fire. The heat was palpable more than a mile away. The burning oil poured as well into the sewers, only to erupt as towering flames from other openings, including home basements. If ever there was hell on earth, this would be it.

About one hundred and fifty firefighters arrived on the scene, some from around Quebec, but also joined by brigades from Maine. Local hospitals prepared for wounded, but none arrived—there were just dead bodies. It took the firefighters almost two days to put out the fires. In all, 47 people died, with 42 bodies recovered. It is assumed that the five other victims were "vaporized" in the intense heat, and nothing will ever be found of them. Over 2000 were evacuated either because of the fire or the deadly fumes that arose because of the fire. The downtown core of the city was decimated, with over 30 buildings gone.

On the fourth day after the accident, Edward Burkhardt, the CEO of Rail World (which owned MMA), came to Lac-Mégantic to survey the damage and received a rough welcome from the townsfolk. Even though Rail World had a poor safety record, compared to other railway companies in the U.S., Burkhardt was quick to blame the train's engineer, suspending him for failing to set the hand brakes properly.

But was the engineer really to blame? Safety officials and legislators on both sides of the Canada–U.S. border keyed in, rightly, on MMA's use of the Single Person Train Operation protocol on their trains. Let's consider some probabilities. Even the most careful experienced train engineer would have some probability of not setting the brakes properly—anyone would. Suppose, for argument's sake, a train engineer has a 95% success rate for correctly implementing the appropriate measures to a train once stopped. That leaves a 5% chance of the train being improperly braked—a 1 in 20 chance of things going dramatically (perhaps even fatally) wrong.

But if we have two engineers on the train, and each independently checks the hand brakes, then the chance of both of them failing to apply the brakes properly jumps from 0.05 down to $0.05 \times 0.05 = 0.0025$, less than 1%. And the results get even more dramatic when we increase the number of engineers checking the cars to three or four, where the probability of all engineers failing drops down to 0.0125% and 0.000625% respectively, miniscule possibilities compared to the original 5% failure rate.

We haven't improved the quality of the checking ability of the engineers, only the number of such engineers, and the decreased risk is dramatic.

Whether the chance of a single engineer improperly braking the train is 5%, 1% or some other number, the principle is still the same—the more independent eyes checking, the more drastically safer the overall system is. I don't know how much blame the one engineer, Tom Harding, should bear. But I do know that a process without built-in independent redundancy is, just by probability, one that is more prone to catastrophic failure. There is always a trade-off in risk versus cost, but the Lac-Mégantic tragedy points out how critical it is to have the issue and its mathematics firmly on the rails.

4.3   Review Exercises

Exercise 4.3.1. If the stated nominal interest rate is 8%, what is the monthly rate? The weekly rate? The daily rate?

Exercise 4.3.2. You are ready to invest $2500. You have a choice of two accounts: Account A pays simple interest at a rate of 6% per year and Account B pays compound interest at a rate of 4% per year. Complete the following table for the amount in each account at the end of each year, just after interest has been paid.

Year	Account A	Account B
1		
2		
3		
4		
5		
10		
20		

Exercise 4.3.3. You take out a loan of $12 000 at 9% per annum, interest compounded monthly. If you don't make any payments until the end of the fourth year, and then pay it back, how much will you return to the lender?

Exercise 4.3.4. You take out a mortgage of $375 000, over 30 years, at a nominal annual interest rate of 6%, compounded monthly. How much are you required to pay at the end of each month, starting with the end of the first month?

Exercise 4.3.5. You turned 25 today and would like invest some money in an annuity that pays you $500 a month from the day you turn 75 until the day you turn 95 (including that day). The nominal annual interest rate is 3%, compounded monthly. How much do you need to put into the account today?

Exercise 4.3.6. How many ways can you choose a password of four letters followed by two numbers and a capital letter? How does this change if all the letters must be different, the two numbers must be different, and the capital letter can't be O or I?

Exercise 4.3.7. Suppose you roll five dice. What is the probability that three dice show one number, the other two another (so that it is a "full house")?

Exercise 4.3.8. Suppose you draw 3 cards from a deck of 52. What is the probability that they are all diamonds or that they are in order (that is, a "straight" of three cards)?

Exercise 4.3.9. If you have no information about the rate of aggression in a certain breed of dog, and you have come across five such dogs, four of which have behaved aggressively, what is your best estimate for the rate of aggression among the breed?

Exercise 4.3.10. A tail gunner in World War II estimated that, on average, only 5% of bombers were lost on any given mission. A tour of duty was 30 missions. What was his probability of making it through the tour of duty? Was he fortunate?

Part II

The Life in Your Math!

MATHEMATICS ISN'T SOME DRY SUBJECT. IT HAS A LIFE OF ITS OWN. And mathematics can make your everyday life better. There may be some decisions weighing on you, and the thought of deciding between your choices may be daunting. There can be a lot at stake, not just for you, but for your family, your friends, significant others, your country and the world! On a lighter note, perhaps you are an artist—professional or amateur, visual artist or musician—who is searching for a great new idea or approach. And perhaps you want nothing more than to enjoy a good thought or a great laugh. You may be surprised that for all of these reasons, mathematics may be just the right thing.

Deciding to Make the Best Decisions

The scientists are quite unanimous—the planet is warming. Ice is melting faster in the Arctic and Antarctic, with dire consequences looming down the road. Over the last hundred years the air and sea temperatures have risen about 1.4°F, with much of the increase over the last 30 years. The increase may not seem like much, but the trend is very worrisome.

Some pundits on the far right say that global warming isn't happening, but I think that they <u>want</u> the answer to be no. NASA's website states that global sea levels have risen almost 7 inches over the last century, with the rate over the last 10 years being almost double that in the last hundred years. NASA also indicates that since 1880, 20 of the warmest years have occurred since 1981.

That brings to mind record-breaking in general. If there was nothing affecting the weather, we would expect some natural variation in average world temperature from year to year, for sure. If we were looking for record-breaking warm years, since 1880, of course 1880 would be such a year, as it was the first. The next year's average temperature would be just as likely to be higher than lower, if chance were the only thing at play, and so the probability that it was record breaking would be 1/2. Likewise, the chance that the third year would be record-breaking would be 1/3 (as the average temperature in the third year would have to be the biggest of the three). And so on. Over the period 1880 to 2001 you would expect $1 + 1/2 + 1/3 + \cdots + 1/122 \approx 5.4$ years. When I look up how many record-breaking years for average temperature there are in this range, I find that there were 14, much, much more than could be expected just by chance. Almost certainly there must be something going on.

The cause of the warming is less certain, but the scientific community has good reason to believe that it is man-made, caused by greenhouse gas emissions formed by burning fossil fuels (and mankind is certainly addicted to the latter). In fact, the Intergovernmental Panel on Climate Change (IPCC), back in 2007, was at least 90% certain of this fact.

227

A worldwide consensus is needed on what can and should be done to remedy the situation.

In 1992 the United Nations Framework Convention on Climate Change (UNFCCC) was formulated in Rio de Janeiro, and over 190 countries have signed on so far. The original document set no strict targets for each country limiting its greenhouse gas emissions, but the subsequent Kyoto Protocol and Cancun agreements did just that, setting a 3.6°F limit to global warming from the temperatures existing before industrialization. But some major players have balked—the U.S. has signed but not ratified the protocol, and Canada has pulled out completely. With all that is at stake—extreme weather, rising ocean waters and increased flooding, drastic changes to animal habitats—why can't the world leaders make the right decision?

Life is full of making decisions. What to eat for breakfast? What university and program to enter? What career to choose? Who to date? How to best land a job? Where to live? And so on. Sometimes the decisions are just for ourselves, sometimes they are in a competitive setting, and sometimes we need to formulate group decisions. We all want to make the best decisions we can, but we often decide—whether what is at stake is very little or a lot—by the seat of our pants, without any process. And later we look back and ask—could we have done better? That is what this chapter is about—finding out how mathematics can help us come to the right decisions. Not necessarily right for everyone, but right for us.

5.1   *Making the Right Choices for You*

While we've talked in the previous chapter about money and risk, neither by itself tells the whole story when it comes to making decisions.

Great Expectations

In the simplest case we are often offered some random process, with the payoff being related to the outcome. For example, perhaps we plan to invest $1000 in a stock for a year. From previous behavior of the stock, you think that

- with probability 0.10 the stock will increase by 30%,

- with probability 0.60 the stock will increase by 8%, and

- with probability 0.30 the stock will lose 15% of its value.

Is it worthwhile to invest? What can you expect to earn (or lose) if you do invest for the year?

The word "expect" in the last question is the key one. A sound way to base your decision would be as follows: if you do invest, then with probability 0.10 your $1000 will increase by $0.3 \times \$1000 = \$300$, with probability 0.60 the increase will be $0.08 \times \$1000 = \$80$, and with probability 0.30 the decrease will be $0.15 \times \$1000 = \$150$, which we'll restate as the *increase* being $-\$150$, using negative numbers to indicate a loss. Then it seems reasonable to weight each possible outcome by its likelihood of happening and add up all the values to get an idea of what will happen overall:

$$0.10 \times \$300 + 0.60 \times \$80 + 0.30 \times (-\$150) = \$30 + \$48 - \$45 = \$33.$$

Here is an example of what you might earn over 20 years, year by year, under this scenario:

$$80, 80, 80, 80, -150, 80, 80, 80, 80, 80, -150, 80, 300, 80, -150, 80,$$
$$80, -150, 80, 80.$$

The total earnings turns out to be $900, which is $\$900/20 = \$45$ per year. In another 20 year period, you might earn

$$80, -150, -160, 80, 80, -150, 80, 80, 80, 80, -150, 300, 300, 80,$$
$$300, 80, -150, 300, 80, 80,$$

which would give, on average, $66.50 per year. The earnings bounce around a bit, but you will find that if the number of years increases dramatically, what you will earn, on average, is indeed

$$0.10 \times \$300 + 0.60 \times \$80 + 0.30 \times (-\$150) = \$30 + \$48 - \$45 = \$33.$$

This value based on the outcomes *and* the probabilities of the outcomes is called **expectation**. More precisely, suppose we have a random process going on, and based on the outcomes we have what is called a **random variable** X, which is simply a variable that takes on a number depending on the outcome of some random process. Here the random variable is how much you will earn/lose. The **expectation** of the random variable X, written as $E(X)$, is the sum of the probability of each outcome times the value of the random variable for the outcome. And by grouping the different values that the random variable can take on, we get a different, useful form—the expectation is also the sum of each value that the random variable can take on times the associated probability of getting that value.

Suppose a random variable X takes on values $x_1, x_2, \ldots, x_k$. The **expectation** of X is

$$E(X) = x_1 \text{Prob}(X = x_1) + x_2 \text{Prob}(X = x_2) + \cdots + x_k \text{Prob}(X = x_k).$$

The expectation is what you would expect to gain/lose, *on average*, if you repeat the random process many times. So in our example, the expectation is $33, what we would gain on average, for investing out $1000, many times.

Here is another simple example. Suppose you can play a game where you pull 1 card from a deck of 52. If the card is a club, you win $2. If the card is a diamond, you lose $1. And if the card is the queen of hearts, you win $20. How much would you expect to win, on average, over repeated plays of the game? How much would you expect to win over 50 games? If you are charged $1 to play, is it worthwhile to play, over the long run? How much would you be willing to play the game repeatedly?

To begin, let's calculate the expected winnings.

- You win $2 if you draw a club, which happens with probability 13/52.

- You lose $1, that is, you win −$1 dollar, if you draw a diamond, which happens also with probability 13/52.

- You win $20 if you draw the queen of hearts, which happens with probability 1/52.

- You win nothing, that is, $0 dollars, if you draw any other card, which happens with probability $(52 - 13 - 13 - 1)/52 = 25/52$.

(We could replace, of course, 13/52 by 1/4, but we'll see that the arithmetic is easier if we don't.) So the expected winnings are

$$2 \cdot \frac{13}{52} + (-1) \cdot \frac{13}{52} + 20 \cdot \frac{1}{52} + 0 \cdot \frac{25}{52} = \frac{33}{52} \approx 0.6346.$$

Thus we would expect to earn, on average, about 63 cents per play of the game. Over 50 games, we would expect to earn $50(0.6346) = 31.73$, that is, 31 dollars and 73 cents (though if we played 50 games, we might earn more or less than this, and certainly never *exactly* this amount, as our winnings or losses are always a whole number of dollars). As the expectation is positive (it is 0.6346), it is worthwhile to play the game. In fact, the expectation tells us exactly how much we should be willing to pay to play the game—we would be willing to pay any amount less than 0.6346, that is, we would be willing to pay 63 cents or less to play the game (as in the long run, we would still earn money) and be unwilling to pay 64 cents or more to play (as then we would lose money in the long run).

Let's take one more example, of a lottery (more about lotteries very soon). The **Powerball lottery** is played across the U.S., drawing in people with the possibility of winning it big. In the basic game, you select five numbers from a set of white balls (W) labeled $1, 2, \ldots, 59$ and one red "Powerball" (PB) from those labeled $1, 2, \ldots, 35$. You can choose your own numbers or have them chosen randomly for you. You win a certain amount of cash, depending on how many balls you match.

The jackpot value differs from game to game, as, if no one wins, the next Powerball jumps by $10 000 000 and a winner of the jackpot has the option of receiving cash or a kind of annuity

You Match	You Win	Probability of Winning
just the PB	$4	0.0180
1 W plus PB	$4	0.0090
2 W plus PB	$7	0.0014
3 W, no PB	$7	0.0028
3 W plus PB	$100	0.00008167
4 W, no PB	$100	0.00005239
4 W plus PB	$10 000	0.000001541
5 W, no PB	$1 000 000	0.0000001940
5 W plus PB	jackpot!	0.000000005707

Table 5.1: Payouts and probabilities for the basic Powerball lottery.

paid over 30 years. If you are a Powerball player, Table 5.1 is a bit depressing—you are about 16 times more likely to be struck by lightning than to win the jackpot!

But is it worthwhile to play the Powerball? After all, the chances of winning are small but the payout is huge! One way to answer this is to find the expected winnings. The random variable in this case is the winnings (it is a number that depends only on the outcome of the random process), and we multiply each winnings by its probability, summing to get the expectation. Suppose that the jackpot will be $250 000 000. Noting that the chances of winning nothing are quite large (0.6136), we can calculate the expected winnings:

$$
\begin{aligned}
E(\text{winnings}) &= 0 \cdot 0.6136 + 4 \cdot 0.0180 + 4 \cdot 0.0090 + 7 \cdot 0.0014 + \\
&\quad 7 \cdot 0.0028 + 100 \cdot 0.00008167 + 100 \cdot 0.00005239 + \\
&\quad 10\,000 \cdot 0.000001541 + 1\,000\,000 \cdot 0.0000001940 + \\
&\quad 250\,000\,000 \cdot 0.000000005707 \\
&= 1.79
\end{aligned}
$$

That is, you can expect to earn, on average, over many plays of the game, only $1.79. Taking into account that you have to pay $2 to play, you lose, on average, $2.00 − $1.79 = $0.21, that is, 21 cents, on each lottery ticket (so over 1000 tickets, you lose, on average, $1000 \cdot $0.21 = $210.00, winning occasionally, but mostly winning nothing). This why the lottery corporation makes so much money, even though they give out big prizes. It's probably better to save your money than invest in the Powerball!

Sometimes indeed we base our decisions on money and how much we expect to gain in the long run. But sometimes money isn't everything.

How You Value What You Value

How much more do you prefer $2 000 000 than $1 000 000— twice as much? Some wouldn't value the former much more than the latter, as a million dollars might be more than enough, while some would value much more twice one million, if they absolutely needed two million right away and one million wouldn't do. And what do we do about ranking things like security and happiness, which don't have dollar amounts? Sometimes you need to make a choice between money and love.

And in terms of risk, would you accept a risk of 10% of dying? It probably depends on more than the risk. If it is a risk jumping from the roof of a three-story building onto the grass below for fun, probably not (I hope!). But if it were to save a loved one in the way of a runaway car, I would think yes! So we make our decisions based on some combination of risk and benefit, and that is what we want to formalize, mathematically, to have a process for making rational decisions. We begin with one approach that has been formulated, but first we need to examine exactly how we value (and ought to value) things.

Mathematicians and economists often attach to each individual a **utility scale** for measuring how much they value items, whether objects or experiences. Formulating such a scale is personalized—each person can value things differently, and there is no concept of an irrational value attached to something. For example, you could value getting a graduate degree much higher than someone else would, and value it differently from how you value say $50 000. We only want our utility scales to satisfy a few rules:

- To every object under question, we need to assign a real number, indicating its worth to us.

- If one object is given a larger number than another, we value it more.

So, for example, Figure 5.1 is a portion of a utility scale I might have. Here I have a utility (or value) of 0 to a quiet day, neither desirable nor undesirable in my books, and a value of 20 to a night out—four times the value I have assigned for finding a $20 bill on the street. And stubbing my toe hard I value at −50—very undesirable, but not as undesirable, I would imagine, as some other events I could imagine (but didn't bother assigning a utility to yet). Now even if I valued losing all my hair as 100, instead of giving a large negative utility, that wouldn't be "irrational" necessarily. My choices are my choices, and the mathematics allows for that.

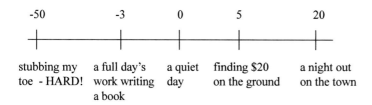

Figure 5.1: Portion of a utility scale.

Before we proceed on how to use utilities in decision making, we pause to talk a little about *how* to form a personal utility scale. It seems a little unsettling to pick one out of thin air—we should have a mechanism for picking out utilities. So here is one way to do it. We start off assigning arbitrarily a utility to some event, say give a utility of 100 to winning a million dollars. Then we imagine playing a lottery with some other event, say winning 2 million dollars, for which we want to assign a utility. The lottery is winning 2 million dollars with probability 1/2 and winning nothing with probability 1/2. Compare it to being given 1 million out right. Do you

Figure 5.2: Lottery of winning $2 million versus getting $1 million outright.

- prefer playing the lottery to accepting the million dollars,

- prefer accepting the million dollars to playing the lottery, or

- express indifference between playing the lottery and being given the million dollars (that is, either is fine—you don't care which you do).

In the third case, if winning 2 million dollars has a utility of u to you (with u unknown so far) then

$$\frac{1}{2} \times u + \frac{1}{2} \times 0 = 100;$$

the right side is the utility of the sure thing (the receiving of one million dollars), while the left is **expected utility** of the lottery, that is, the sum of each utility times its probability of happening. Solving for u, we find that $(1/2)u = 100$, that is, $u = 2 \times 100 = 200$, so *your* utility of receiving 2 million dollars is 200.

If you prefer the lottery to the million dollars, then the utility of 2 million dollars satisfies

$$\frac{1}{2} \times u + \frac{1}{2} \times 0 > 100,$$

so again by solving you find that $u > 200$, that is, receiving 2 million dollars is worth more than twice the utility of getting 1 million dollars. Our next step is to change the probabilities and move *down* the chances of getting the 2 million dollars. So you be asked to compare playing a lottery of winning 2 million dollars say with probability 1/4 and winning nothing with probability 3/4, versus getting a million dollars for certain. If you are indifferent between the lottery and the outright payment of 1 million dollars, then

$$\frac{1}{4} \times u + \frac{3}{4} \times 0 = 100,$$

so u is 400, that is, receiving 2 million dollars has utility 400 to you. If you prefer the lottery, you move the probability of receiving the 2 million dollars in the lottery even lower, and proceed; if you prefer the certain payment, you move the probability of receiving the 2 million dollars in the lottery higher (but still lower than 1/2) and proceed.

You do a similar process (but move the probability of winning 2 million *up*) if you prefer the million dollar certain payout to the

Figure 5.3: Second lottery of winning $2 million versus getting $1 million outright.

lottery. Eventually, you should settle on a lottery for which you are indifferent between it and the million dollar payoff, and then you calculate your utility u from setting the expected utility of the lottery equal to 100, the utility of 1 million dollars.

For undesirable outcomes, you can repeat the process, setting say at -100 the utility of some bad event, and run lotteries with the other undesirable outcomes with the set one (though it may seem strange to compare a lottery where you lose a finger to the certainty of losing $50 000, but this is exactly the type of lottery you invest in with insurance).

The process can be a long one, even for determining a single utility, let alone a bunch of utilities, but as long as you have a feeling for probabilities, it does give you a method for assigning probabilities.

Deciding with Maximum Expected Utility

Now let's use utilities to make decisions. Our approach is to attach utilities to all of the possible outcomes and then weight them by the chances of them occurring (so that the utilities are a random variable that depends on the outcomes). The decision process of **maximizing expected utility** (MEU) is to take the *largest* of all the expected utilities available.

I think this calls for an example! Let's suppose you have a choice—stay in and study tonight or go out and party. If you stay in and study, there is a 90% chance you'll do well on your math midterm tomorrow, but if you go party, the chance that you'll do well on the midterm drops down to 50%. Now what about fun? If you stay home, there is only a 30% chance you'll have any fun (you do enjoy math a bit, but you know if you stay in, a friend may drop over, and you would enjoy that). On the other hand, if you go out, there is an 85% chance you'll enjoy the evening. The fun and score on your midterm are dependent on whether you stay home or not, but given one of the choices, the fun and midterm scores are independent of one another. Figure 5.4 is a **decision tree** that shows the choices you have, along with the probabilities of various outcomes occurring (diamonds indicate choices, circles indicate randomness). Note that the probability on a branch is a conditional one, based on

the intermediate outcomes along a branch.

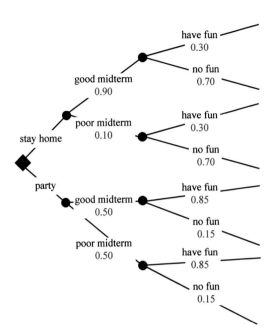

We need to add in utilities for the various outcomes. In Figure 5.5, at the end of each branch in the decision tree I have attached some utilities. For example, if I stayed at home, had a good midterm and had fun, I assign this a utility of 80, while if I partied, had a poor midterm and had no fun on top of that, I give this a utility of −30.

Each branch leads to an outcome. What is the probability of a branch? Once you make a choice, the probabilities are independent along a branch (as mentioned earlier, whether you have fun or not is independent of whether you score well on the midterm, once you have made your decision). So the probability along a branch is simply the product of the probability along the edges of the branch. Once you have the probabilities, you multiply along by the utilities and sum to get the expected utility for each decision. The expected utility for staying at home is

$$0.90 \cdot 0.30 \cdot 80 + 0.90 \cdot 0.70 \cdot 60 + 0.10 \cdot 0.30 \cdot 30 + 0.10 \cdot 0.70 \cdot 0 = 60.30,$$

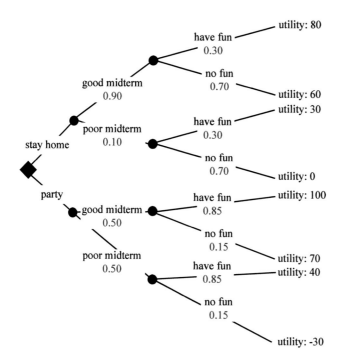

while the expected utility for partying is

$$0.50 \cdot 0.85 \cdot 100 + 0.50 \cdot 0.15 \cdot 70 + 0.50 \cdot 0.85 \cdot 40 + 0.50 \cdot 0.15 \cdot (-30) = 62.50.$$

So comparing these two and choosing the largest expected utility, you decide to party (you might have thought that I would have chosen the situation so that the optimal thing to do was to stay home and study, but I let the numbers do the talking!).

The method of maximizing expected utility is one that has been touted regularly as "the" way to make a rational decision, but a good case can only be made for it when you repeat often the same decision scenario—in that case, over the long run, you will get the "most" utility if you always choose by MEU. But what if you have a single decision to make? It is not so clear that you should use MEU (though it is often put forward as what you should do). For example, suppose you have a decision to make, to choose A or B, with the decision tree in Figure 5.6 (the utilities are actually dollar amounts). Here we find that decision A has expected utility

$$0.90 \cdot 1000 + 0.10 \cdot 0 = 900,$$

while B has expected utility

$$0.001 \cdot 1\,000\,000 + 0.999 \cdot 10 = 1009.99,$$

so on the basis of maximum expected utility, we should definitely choose B over A, and indeed, in the long run, if we had
to choose over and over again, B would bring you much greater
results than A.

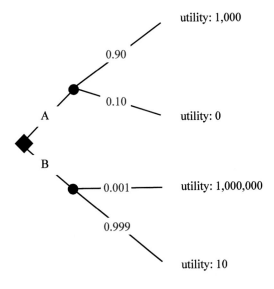

utility: 1,000

0.90

0.10

A

utility: 0

B

0.001 ——— utility: 1,000,000

0.999

utility: 10

Figure 5.6: MEU is not always
the way to go!

Having Enough

But what if you only get to choose between A and B once?
Then I think you could make a very good case that you should
choose A, as you have a large probability (90%) of getting 900,
while only a very tiny, 0.1%, chance of getting more than 10
in the second. On a one-off decision, I would think A beats B,
hands down.

But even if you repeat, there are other accepted ways to
make "rational" decisions. One that has been proposed is called

satisficing, a mix of *satisfying* and *sufficing*, where you don't select the optimal solution, but simply one that meets a certain level of need. For example, if you were given the choice between taking $100 000 and a lottery with a 50% chance of winning $1 000 000, if $100 000 were sufficient for your needs, why not select the $100 000, even though the lottery gives you much better expected earnings?

> *Life Lesson:* Whether you are a maximizer or a satisficer, the key is to spend some time thinking about your decision process. At the very least, it would be a good decision to do that!

Exercises

Exercise 5.1.1. What is the expected number of pips shown on a rolled die?

Exercise 5.1.2. Suppose you are offered to play a game repeatedly where you roll a die where you win 2 dollars if you roll an even number, and lose 1 dollar if you roll an odd number. Would you be willing to pay 50 cents to play the game each time? How much would you expect to win if you played the game 25 times?

Exercise 5.1.3. Suppose you are offered to play a game repeatedly where you roll a die where you win 2 dollars if you roll an even number, and lose 1 dollar if you roll an odd number. Would you be willing to pay $1.50 to play the game each time? How much would you expect to win if you played the game 25 times?

Exercise 5.1.4. Suppose you are offered to play a game repeatedly where you roll a die where you win, in dollars, the number of pips. Would you be willing to pay $3 to play the game each time? How much would you expect to win if you played the game 100 times?

Exercise 5.1.5. Suppose you flip a coin twice, and you win $5 dollars for each heads you toss (so, for example, if you toss tails and then heads, you win $5 dollars, as you tossed one head). Would you play $3 dollars to play the game repeatedly? How much would you expect to win/lose, on average, per game?

Exercise 5.1.6. Suppose you repeatedly play a game where you pull a single card from a deck of 52 cards. If the card is a heart but not a face card, you win 1 dollar. If the card is a heart that is a face card you win 5 dollars. And if the card is the ace of spades, you win 10 dollars. What is the most you would be willing to pay to repeatedly play the game?

For Exercises 5.1.7 to 5.1.11, suppose that you plan to invest $10 000 in a stock for a year. From previous behavior of the stock, you think that

- with probability 0.20 the stock will increase by 40%,

- with probability 0.40 the stock will increase by 5%, and

- with probability 0.40 the stock will lose 10% of its value.

Exercise 5.1.7. Simulate a number of years of investments of $10 000 in the following way. Take 10 cards, ace through 10 (the suits do not matter, just one of each denomination), from a deck of 52. Shuffle them several times and then take the top card:

- If the card is an ace or a 2, then the stock has increased by 40%.

- If the card is a 3, 4, 5 or 6, then the stock has increased by 5%.

- If the card is a 7 or higher, then the stock has decreased by 10%.

(a) Explain why this method properly simulates the stock behavior.
(b) Simulate the stock behavior 20 times, as above, and write out how much you have won or lost each time.
(c) Calculate the average amount earned/lost in part (b).
Exercise 5.1.8. What can you expect to earn (or lose) if you do invest for the year? How does this compare to the answer you got for the previous simulation, part (c)?
Exercise 5.1.9. Is it worthwhile to invest?
Exercise 5.1.10. What would be the most you would be willing to pay a stock broker each year to invest your money repeatedly, if you expect to earn, on average, 5% on your investments?
Exercise 5.1.11. What would be the most you would be willing to pay a stock broker each year to invest your money repeatedly, if you expect to earn, on average, 7% on your investments?

Exercise 5.1.12. Confirm that in the Powerball lottery, the probability of matching just the Powerball is 0.0180. (Hint: note that if you match just the Powerball, then you don't match any of the white balls.)
Exercise 5.1.13. Confirm that in the Powerball lottery, the probability of matching three white ball and not the Powerball is 0.0028.
Exercise 5.1.14. Confirm that in the Powerball lottery, the probability of getting the jackpot is 0.000000005707.

In the *Atlantic 49* lottery, six numbers are chosen randomly from 1 to 49, inclusive, along with a seventh bonus number (from the same set of numbers). Use this information to answer Exercises 5.1.15 to 5.1.18.

Exercise 5.1.15. Explain why the probability of matching all six (regular) numbers is $1/13\,983\,816$.

Exercise 5.1.16. Complete the following table:

You Match	You Win	Probability of Winning
2 out of 6 plus the bonus number	$4	
3 out of 6	$6	
4 out of 6	$64	
5 out of 6	$649	
5 out of 6 plus the bonus number	$64\,900	$1/2\,330\,636$
6 out of 6	$1\,000\,000	$1/13\,983\,816$

Exercise 5.1.17. The cost to play Atlantic 49 is one dollar. Based on the payouts in the previous exercise, what are your expected winnings?

Exercise 5.1.18. Suppose instead of just winning $1\,000\,000$ when you get 6 out of 6 numbers, you win a jackpot. How big does the jackpot have to be for playing to be worthwhile, that is, have positive expected winnings?

Exercise 5.1.19. For the following events, assign "reasonable" personal utilities, on a scale from -100 to 100 (do not bother to use lotteries for this question): taking a walk, cooking dinner, cleaning up after dinner, playing soccer, reading a book, writing an essay, doing math homework, listening to music. Then put the utilities you assigned in the previous exercise on a number line.

Exercise 5.1.20. Suppose a person has assigned a utility of 100 to 1000. If she is indifferent between getting 1000 outright or a lottery where she can win $100\,000$ with probability 0.10 and nothing with probability 0.90, what utility should she assign to $100\,000$?

Exercise 5.1.21. Suppose a person has assigned a utility of 100 to 1000. If she is indifferent between getting 1000 outright or a lottery where she can win 1500 with probability 0.20 and nothing with probability 0.80, what utility should she assign to 1500?

Exercise 5.1.22. Imagine you are offered a choice between a lottery where you have a 50% chance of winning 200 (and a 50% chance of winning nothing), versus receiving 100 outright. Which do you choose?

Exercise 5.1.23. Run lotteries, as described in this chapter, to find your utility for 200, given you have assigned a utility of 10 for receiving 100.

Exercise 5.1.24. If, when you assign utilities to money, whenever you have assigned a utility of x to some dollar amount y, you always assign a utility of *less than* $2x$ for $2y$, are you a risk taker or a risk avoider? Explain your answer.

Exercise 5.1.25. Suppose the staying at home versus partying example in this chapter has different utilities:

- If you stayed at home, had a good midterm and had fun, you assign this a utility of 90.

- If you stayed at home, had a good midterm and had no fun, you assign this a utility of 75.

- If you stayed at home, had a poor midterm and had fun, you assign this a utility of 40.

- If you stayed at home, had a poor midterm and had no fun, you assign this a utility of 10.

- If you partied, had a good midterm and had fun, you assign this a utility of 85.

- If you partied, had a good midterm and had no fun, you assign this a utility of 55.

- If you partied, had a poor midterm and had fun, you assign this a utility of 30.

- If you partied, had a poor midterm and had no fun, you assign this a utility of −50.

With the same probabilities as in the example, find the expected utilities of each decision.

Exercise 5.1.26. With the same utilities as in the previous example, what decision would you make using MEU?

Exercise 5.1.27. Now suppose the utilities are as in Figure 5.5, but the probabilities have changed:

- If you stay in and study, there is a 95% chance you'll do well on your math midterm to-morrow, but if you go party, the chance that you'll do well on the midterm drops down to 40%.

- If you stay home, there is a 50% chance you'll have fun, while if you go out, there is an 80% chance you'll enjoy the evening.

Again, the fun and score on your midterm are dependent on whether you stay home or not, but given one of the two, they are independent of one another.

Find the expected utilities of each decision.

Exercise 5.1.28. With the probabilities (and utilities) in the previous exercise, what decision would you come to via MEU?

Exercise 5.1.29. Suppose that in the stay home-partying example, you had a third choice—to go to a midterm tutorial. If you go to the tutorial, the probability that you have a good midterm goes up to 0.95, while the probability you have fun is 0.60 (as some of your friends will be there too). Once again, the fun and score on your midterm are dependent on whether you stay home, party or go to the tutorial, but given one of the three, they are independent of one another. You assign utilities as follows:

- If you go to the tutorial, had a good midterm and had fun, you assign this a utility of 100.

- If you go to the tutorial, had a good midterm and had no fun, you assign this a utility of 85.

- If you go to the tutorial, had a poor midterm and had fun, you assign this a utility of 50.

- If you go to the tutorial, had a poor midterm and had no fun, you assign this a utility of −40.

With the probabilities and utilities for the stay at home and partying choices as in the example, draw up an appropriate decision tree.

Exercise 5.1.30. For the previous exercise, find the expected utilities of each decision. What decision would you come to via MEU?

Exercise 5.1.31. Suppose you have a choice between receiving $100, a lottery where with probability 0.50 you get $300 and get nothing otherwise, and a third lottery where you get $40 with probability 0.75 and $800 with probability 0.25. Suppose also that your utilities for an amount of money are equal to its value.

(a) If you use MEU, which do you choose?

(b) If you are a satisficer where $75 is enough for you, which do you choose?

5.2   *Game Theory—Coming out on Top*

Sometimes your decisions are only based on your wants and needs, but often there can be others involved, with their own desires and requirements. In this section we'll be interested in maximizing what you get, in a competitive environment, when each player has choices to make. We call such situations **games**, even though the outcomes for each "player" may be important. Our particular focus is on **two-person games**, that is, games where there are two opponents, each trying to get the most for themselves. Some of the games (like poker) are **zero-sum**, in that what one person wins, his or her opponent loses, but others are not. Finally, some games are played with **perfect information**, that is, each player knows exactly what has transpired in the game, the moves that have been made already in the game—examples are tic-tac-toe, chess and checkers—but some games are not (including those that involve an element of chance, such as blackjack).

What we are always after is a "best" way to play the game, in the sense of maximizing what we get. (This is indeed rather uncooperative; in the next section we'll talk about coming to joint group decisions.) Such a way to play, which specifies moves according to previous choices made by both players, is called a **strategy**, and the holy grail is to find a **winning strategy**, if it exists—one that guarantees a win. Finding a winning strategy often requires thought, but it is well worth your trouble, and you

learn a lot, both about games and life in general, in the process.
Throughout, when a game is introduced, find a friend and play
it a few times, and see if you can figure out good/winning
strategies (you'll get much more out of the discussions if you
play along!).

Strategy 1: Working Backwards

Let's look at a simple game called **Nim**, a game you probably
haven't heard of, but an easy one to explain (even a child can
play it!). The game starts off with a number of coins in front of
you and an opponent. One of you chooses to go first. Then play
alternates, with each player choosing to remove 1, 2, 3, 4 or 5
coins on his or her turn. The player that wins is the one who
takes the last coin.

OK, let's try the game out, you versus me. Suppose there are
26 coins in front of us.

Let's suppose we have decided for me to go first. I decide to take
two coins away:

Suppose now that you decide to take three coins:

Now I choose three coins as well:

From the 18 coins left, you decide to take 5:

I reply by selecting one coin:

You choose 4 coins from the 12 remaining:

I take two coins, leaving you with six:

Now you begin to sweat a bit and realize that you have lost. No matter how many coins you now take—1, 2, 3, 4 or 5—I can take the rest and win!

So what gives—did I just get lucky, or did I have a plan? The answer is the latter. Let's see if we can deduce what the winning strategy is. I think you realized that once I left you with six coins, I was guaranteed to win—no matter how many you took, from one to five, I could take the remainder. Now how can I be sure to leave you with six? If I leave you with 12 the previous time, I can always make sure I leave you with 6 the next time around (and win), as if you take

- 1 coin, I take 5,

- 2 coins, I take 4,

- 3 coins, I take 3,

- 4 coins, I take 2, and

- 5 coins, I take 1.

So I see if I can leave you with 12 coins, I can guarantee a win for myself. Likewise, if I leave you with 18 coins the time before, I can make sure to leave you with 12 the next time, and win, and so on. *As long as I leave you with a multiple of 6 coins (like 6, 12, 18, 24, and so on), I can make sure I win*, so this is my winning strategy. So in the game we just played, as the number of coins was 26, my first move was to bring it down to 24, a multiple of 6, by removing 2 coins, and from then on, I remove just enough coins to always leave you with a multiple of 6.

Can I always do this? Well, if I go first, I can, *unless we start the game with a multiple of 6 number of coins*. In this case, no matter how many coins I take away, you can always leave me with a multiple of 6 number of coins, and so you win! So the second player to move can guarantee a win for himself or herself just in the case where the number of coins at the start is a multiple of 6. The fascinating part of figuring out the strategy is the following principle, which you can file away in your mind for use in real-life problems:

> *Life Lesson:* When trying to strategize, sometimes it helps not to work forward, from where you start, but to work backwards, from where you want to end.

So we have completely described the winning for Nim—all that it depends on is the number of coins n at the start (if n is not a multiple of 6, the first player can ensure a win, and if n is a multiple of 6, the second player can guarantee a win). Here, in Nim, we found the winning strategy, with a bit of work, but the strategy itself was not so hard to describe.

Strategy 2: Casework

Let's now turn to a game I'm sure you played when you were young—tic-tac-toe. The rules are pretty straightforward. There are two players—one who plays "X" and the other "O". Player X (or just X) goes first and places an X in a 3 × 3 set of squares, and Player O (or just O) responds, placing an O somewhere else. The game ends when one player gets three of his letters in a row (horizontal, vertical or diagonal), drawing a line through the winning letters. The game ends in a tie (or "draw") if all nine squares are filled with no "X" lines or "O" lines. This is a game of perfect information—everyone knows the state of the board and everyone's choices—and ends after at most nine moves (as there are only nine places to put a letter).

You probably played the game many times, but have you ever wondered about a winning strategy? The game has a lot of moves available to both players, and so it looks too difficult to analyze. But let's wade in at the shallow end of the pool and see how far we get.

There are nine possible first moves for X, but really there are only three essentially different ones—playing in the center square, a corner square or an outside middle square. We have cut down the opening moves to just three from nine, using the *symmetry* of the board, seeing that any other move would be essentially the same as one of the three shown.

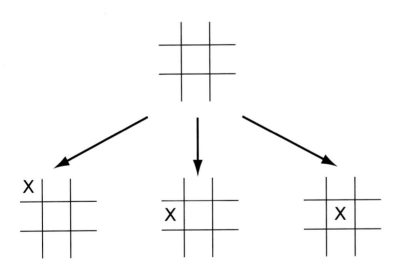

Let's take one of the cases, say the third, with X going first in the center. Again, by symmetry, O has only two different moves to choose from:

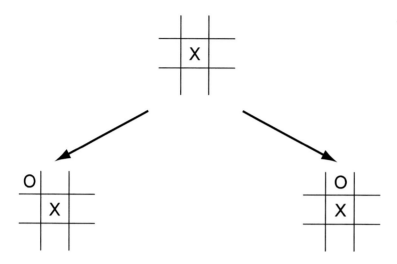

We need to describe how X might respond in each of these two cases. For the one on the left (with O in a corner), there are essentially four different possible responses for X:

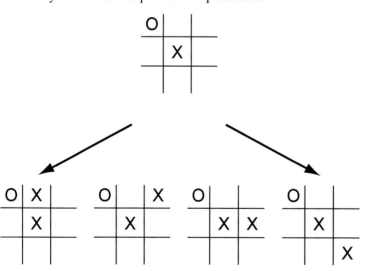

For the first one, O must respond below the two X's, or X will win on his next move. After this, X shouldn't play in the upper right corner or the square below it, as then O will play in the lower left, having <u>two</u> ways to complete a line, only one of which X can block. So there are four choices for X left to play, and you can check (with a bit of work on the side) that in each case, no matter how X and O play afterwards, the game ends in a tie). Similar types of arguments, in each of the other three cases (do them!) will show that the best X can do in each case is tie.

But these four cases are only the ones where X starts in the center. We need to examine the cases where X starts in a corner or the middle of the outside. There are lots of cases to be handled individually! I have done so, and I checked that in all cases, the best that O can do is force a tie. Figure 5.8 shows, for the possible opening moves for both X and O, what is the best that can be achieved with each person playing optimally.

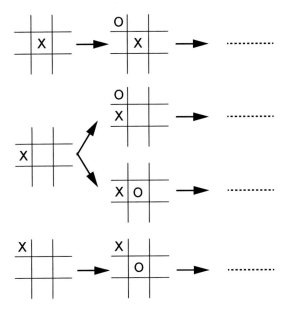

Figure 5.7: O's strategy for tic-tac-toe (to ensure the game ends in a tie). Note that O's strategy depends on X's opening move, and describes what to do in each case.

Thus if X plays the center to begin with, then O can ensure a tie (by playing a corner), which is the best outcome from O's point of view. If X opens with a middle square on the outside, then O can ensure that the game ends in a tie. Finally, if X plays a corner, O can still ensure a tie by playing the center. So in any event, no matter what X does, O can guarantee a tie, and since the casework shows that this is the best that O can achieve (other moves can only result in a loss for O), the outcome for tic-tac-toe is a tie, with O's best strategy (depending on each of

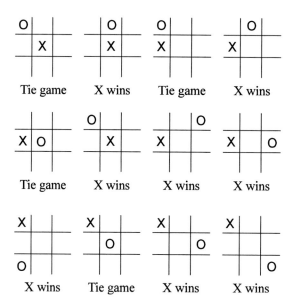

Figure 5.8: Outcomes for tic-tac-toe, based on the opening two moves.

the essentially three different opening moves for X) shown in Figure 5.7 (the strategy is not fully listed, only the best opening move for O, but can be reconstructed from the arguments we used to find the result of each of the openings in Figure 5.8).

That was a big effort, but sometimes that is exactly what needs to be done. The math allowed us, through the use of symmetry, to cut down drastically on the work.

> *Life Lesson:* Sometimes casework is unavoidable when solving a problem; the key is to use whatever symmetry is inherent to cut down on the number of cases you need to handle.

Strategy 3: Tweedledum and Tweedledee

In relationships, it almost always a bad idea to copy what your partner does, especially if it is in the midst of an argument. But when you are playing a game to win, sometimes mirroring what your opponent does is *exactly* the right thing to do! Such an approach is called a **Tweedledum and Tweedledee strategy**,

Figure 5.9: Alice, Tweedledum and Tweedledee.

named after two characters in Lewis Carroll's 1871 book, *Through the Looking Glass, and What Alice Found There*. Tweedledum and Tweedledee were two round little brothers who complimented each other endlessly back and forth, and have been portrayed in movies as identical twins who mirror each other's gestures and words. Hence the connection to the strategy.

Figure 5.10: Charles Lutwidge Dodgson (1832–1898) wrote under the pen name "Lewis Carroll." His day job, though, was as a mathematician, specializing in logic!

Let's see the strategy at work. Consider the following game, played with coins like Nim, but with a different wrinkle. There are two stacks of coins, and you play against an opponent. On a move you are allowed to select as many coins as you like (though you must select at least one), but *just from one stack*. Like Nim, the person who takes the last coin wins. Let's do an example. We'll start with two stacks, one with 11 coins and the other with 8 coins. Let the first player be Andrea, and the second player be Ben.

It would be a poor move for the first player, Andrea, to take all of the coins in one stack, as then the other player will take the whole other stack and win. So suppose Andrea takes four coins from the left stack, leaving seven.

The second player to move, Ben, now decides to take two coins also from the left stack.

Andrea decides to take three coins from the stack on the right.

Ben now elects to take four coins from the right stack, leaving just one there.

That was a mistake! Andrea seizes the opportunity, and selects four coins from the left stack, leaving two stacks, each with one coin. No matter which coin Ben takes, Andrea takes the other and wins!

So can we figure out what the winning strategy is? Let's think about it first when both stacks have the same number of coins. In this case, here is a Tweedledum and Tweedledee strategy for the second player: copy whatever the first player does, but in the opposite stack. So if the first player removes five coins from the right stack, the second player responds by removing five coins from the *left* stack. This always leaves the stacks with the same number of coins, as they start with the same number of coins. So the second player always has a move—he can always match the first player's move. We see that the Tweedledum and Tweedledee strategy for the second player always ensures he never gets stuck, and as the game must end at some point, he can't lose. The first player must be the loser.

What about if the stacks contain different numbers of coins? In that case, the first player removes enough coins from the larger stack to make the two stacks equal. At that point, the game is turned over to the second player, and it is the first player who takes on the Tweedledum and Tweedledee strategy, copying the *second* player's moves, in the opposite stack, from now on. The first player has, by moving directly to two equal stacks, converted the game into an equal stacks game with the second player making the *first* move. This turns the tables and allows the first player to win.

I think we have learned a lot from playing this game.

Life Lesson: Sometimes the best strategy is to mimic what your opponent does. And sometimes you can, in a move, turn a game around so that you and your opponent play reverse roles.

Strategy 4: Give a Little, Gain a Lot

There are some games where the players gather up points, with the winner determined by who has gathered up the most points along the way. Perhaps you played one such game, **dots and boxes**, when you were younger. The game starts with a rectangular array of dots:

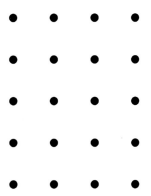

You and I will play each other. We alternate moves, joining a pair of dots next to one another, either horizontally or vertically. The object is complete a box (which is a square of neighboring points); once you do, you get to put your initial in it (J for me, Y for you) and make another move. The game ends when all the boxes have an initial in them, with the winner being the one who has the most boxes.

Let me go first.

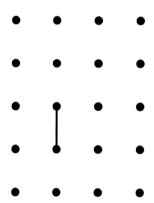

You respond by choosing another pair of adjacent points to join, as follows.

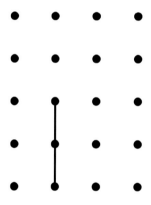

It's my turn again.

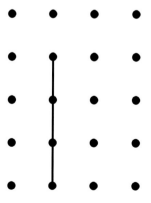

Now it's your turn.

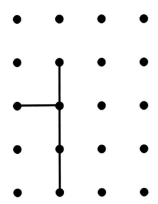

I'll join the pair at the bottom left.

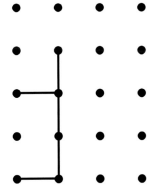

You choose another line.

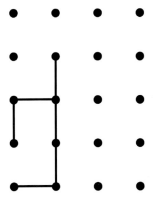

Now I can fill in a box, and I put my initial (J) in it.

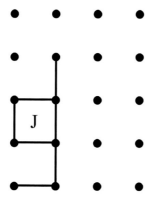

I get another move after filling in a box, and there is now a second box for me to complete.

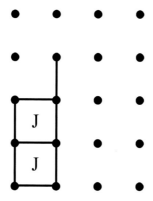

I still have to make another move (after filling in a box), and do so.

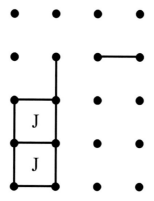

The game continues, with you and me alternating moves, until there are no more possible lines to add. At that point, we count up boxes and see who has more.

So now we have a handle on the game (if we didn't previously). One player, either the first or the second, has a winning strategy—in any game with perfect information which ends with one player winning after some finite numbers of steps, one player must have a winning strategy. But this game is so complicated that it is unknown who has the winning strategy, let alone what such a strategy might be.

But that is not what I'm after here. I want to show you something truly remarkable. Suppose you and I are playing again, on the same board, and part way through, the board looks as follows:

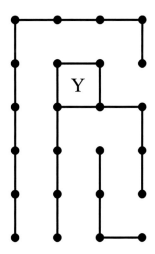

Toward the end of a game, the board typically breaks up into long chains of potential boxes, only needing one more line for the other person to be able to complete the string. Suppose that it is your turn to move. You can either fill in a line and give me the six boxes in the lower right, or fill in a line in the long outside chain and give me eight boxes. What do you do? Of course, you give me the six boxes, knowing you get the eight and win! So you fill in a line like the one shown and wait for my move.

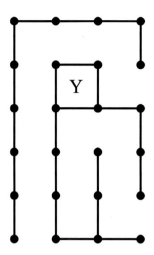

You can just anticipate my filling in the six boxes, with the look on my face as I have to give you the remaining eight. But hold on—do I need to take all six? Isn't that a silly question? Of course, I should take as many boxes as I can, right? But no, I'll show uncommonly good restraint and fill in only four of the six boxes that I can.

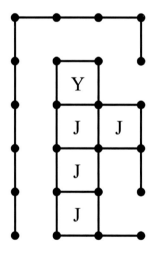

I then stop short of taking the next two boxes, which I am entitled to do, and draw in a line that allows you to take them. How generous of me!

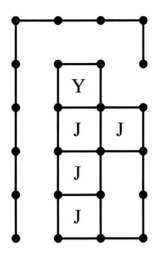

But I am not as altruistic as I seem. You are free to take the two boxes, in fact, with one line, filling in your initials in them.

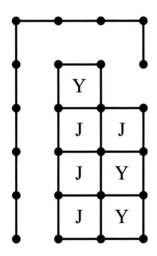

But you feel hollow as I have turned the tables on you—now *you* need to add a line to the long chain.

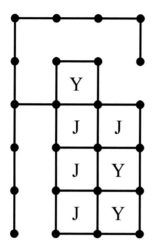

Of course, now I thank you for your move, and I take all of the remaining boxes in the chain, winning 12 boxes to 3. What a disappointment for you—you thought I had made a mistake, not taking what I could at the time, but what I was showing was better judgment, realizing that it is not always in my best interest to take all I can get at the time.

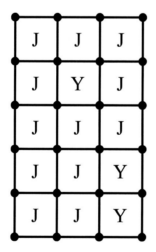

The moral of the story is:

Life Lesson: Sometimes it is better to hold off on taking all you can at the time, even when you are trying to win. Less now may mean more later!

Strategy 5: Sometimes It's Best to Mix It Up

All of the games we have seen so far have no random elements to them. But even if there is no whiff of randomness in the game, sometimes it is best to mix things up. For example, consider the classic two-person game of **rock–paper–scissors**. In case you have forgotten, here are the rules: you and an opponent, at the end of saying "one-two-three," each throws out a hand, showing one of three outcomes: rock (a fist), paper (a flat hand) or scissors (two fingers out). Then

- rock crushes scissors,

- paper covers rock, and

- scissors cuts paper.

A lot of people seem to love to play this simple game—there are even tournaments, both national and international. Note that each of the two players, which we'll call Player I and Player II, has three strategies—R (rock), P (paper) or S (scissors). We'll let R_1, P_1 and S_1 be the corresponding strategies for Player I, and R_2, P_2 and S_2 be the corresponding strategies for Player II. When we can list the strategies for both players, we can form what is called a **game matrix**, which is a table with Player I's strategies listed along the left edge, Player II's across the top, and the entries correspond to what each player wins or loses if they choose the respective strategies. The game matrix is a handy way to keep track of the payoffs for the game. For example, suppose that a player gets a dollar (1) if he or she wins in rock–paper–scissors and loses a dollar if he or she loses (if they tie, no money changes hands). Putting the winnings into ordered pairs, where

Player I's winnings come first and Player II's winnings come second in the table, we get the game matrix shown in Table 5.2.

	R_2	P_2	S_2
R_1	$(0,0)$	$(-1,1)$	$(1,-1)$
P_1	$(1,-1)$	$(0,0)$	$(-1,1)$
S_1	$(-1,1)$	$(1,-1)$	$(0,0)$

Table 5.2: Game matrix for rock–paper–scissors.

So, for example, the entry in the row labeled R_1 and column labeled P_2 is $(-1,1)$ since if Player I plays rock and Player II plays paper, then Player I loses a dollar (-1) and Player II wins a dollar (1). Of course, this game is a zero-sum game—what Player I wins Player II loses, and vice versa—so we could leave out the outcomes for Player II, realizing they are just the negative of the outcomes for Player I:

	R_2	P_2	S_2
R_1	0	-1	1
P_1	1	0	-1
S_1	-1	1	0

So what is the best strategy for each player? It is clear that if Player I knows Player II's strategy, Player I can always choose her strategy to win, and vice versa, so each player should mix it up. Moreover, the choices for each player should not show any discernible pattern that the other player can pick up on, or even any inclinations. With some work, one can show that the best each player can do, over many plays of the game, is to randomly play one of the three strategies available equally likely—with probability 1/3 each—with the expectation of breaking even over the long haul.

So what do we learn here?

Life Lesson: Sometimes when you are competing over the long run, it may be best to randomly choose your strategy to keep your opponent guessing (and unable to figure out any pattern in your play that he or she can use to his or her advantage).

The Problem with Games

Game matrices are useful ways to summarize the strategies and their outcomes for games, and you can sometimes do your reasoning with them. The actual way the game is played is not important once the possible strategies have been found—we only need to choose the best one from among them.

Suppose we play a game where each of us chooses a number from the numbers 1, 3 and 4 to shout out at the same time. If the two numbers add to exactly 6, you win 5 dollars from me. If they add to 7, no one wins anything. In all other cases you pay me 1 dollar. For example, if you shout out "1" and I shout out "4," the sum is 5, not 6, so I win a dollar from you. On the other hand, if we both shout out "3," the sum is 6 and I pay you 5 dollars. Here is the game matrix:

	I shout out 1	I shout out 3	I shout out 4
you shout out 1	−1	−1	−1
you shout out 3	−1	5	0
you shout out 4	−1	0	−1

What should you do? If you look at the game matrix, you see that your strategy to yell out "3" is always better for you, no matter what I do—that corresponds to the fact that in the game matrix, each value in the row for "you shout out 3" is at least as good as, and sometimes better than, any other choice in the column. That means that your second strategy **dominates** the others, and so you can remove the other **dominated** strategies as options—why would you choose a strategy that is *never* as good as another one?

	I shout out 1	I shout out 3	I shout out 4
you shout out 3	−1	5	0

So you have settled on your second strategy, and I realizing this, choose my first strategy—it forces you to pay me a dollar, which is better than my other two strategies, which has me either paying you 5 dollars or winning nothing. By reasoning through, we have found that the optimal strategy for both of us is for you to shout out 3, me to shout out 1. No matter how many times we play the game, there is no reason for us to play any differently.

Removing dominated strategies for either player can help reduce our choices, but often there aren't any dominated strategies, and trying to reason out the best strategy is difficult, if not impossible. We'll talk about three such two-person games—Chicken, the Battle of the Sexes and the Prisoner's Dilemma—that come up so often in real life, they have their special names. Each game is very simple—each player has only two strategies—but playing optimally is very difficult!

The Game of Chicken

I remember a game called *Chicken* that I played when I was younger, with a sibling or a friend, where we would each take one end of a rubber band and pull. The one who let go was deemed the loser. Anticipation built to see who would "chicken out," and the winner, nursing a sore hand, felt like he or she came out on top.

You might think that such games are left behind when you grow up and get smarter, but such is not the case. I was traveling in a big city a while ago when I came across two drivers, both stopped in the middle of an intersection, over some sort of disagreement. They both put their cars in park and jumped out. One was a young guy, maybe 20 years old, and the other was an older man, perhaps 65. They were in each other's faces, staring. The old man even grabbed the cords from the ear bud headphones in the young guy's ears and yanked! That only added to the scene.

Each of the combatants had one of two choices—confront or back off. Of course, the payoffs for each player aren't money, but intangibles like self-esteem, health, and so on. This is a perfect opportunity to assign utilities (as discussed in the previous section) for the values of the outcomes to each player. I made some up, depending on how I think the "players" would feel after their joint decisions. The utilities in Table 5.3 show in each box an ordered pair, with the old guy's utility first, followed by the young guy's utility (that is our convention—in these game matrices, the first number corresponds to the row, the second

to the column).What is important here is not the exact numbers
chosen so much as the rankings of the utilities—how the values
of the outcomes sit with respect to one another for each person.

	young guy confronts	young guy backs off
old guy confronts	$(-200, -200)$	$(100, -100)$
old guy backs off	$(-100, 100)$	$(50, 50)$

Table 5.3: Game of Chicken.

The key thing is that it is worse for both if they confront each
other—serious (stupid) injury is likely imminent. It is jointly
best for both to back off and walk away. But if they both realize
this, each of them will confront, based on the assumption that
the other will back off. So we see that the situation is *unstable*—
the best joint decision is to back down, but if you assume your
opponent realizes this, then it is in your best interest to confront.
Then the end result? Confrontation—the worst joint outcome,
I think. And there is little you can do about it. I chose to drive
off without seeing how the game of Chicken ended with the two
drivers.

The Battle of the Sexes

Here is another class of games for which simple self-interest
also gets the better of you. Suppose there is a couple deciding
what to do on vacation—either to go on a tour of Asia (the wife's
preference) or to go on a tour of U.S. ballparks (the husband's
preference). They are fairly adamant about their positions, and
Table 5.4 is a game matrix for their choices and utilities.

	husband chooses BP	husband chooses Asia
wife chooses BP	$(10, 100)$	$(-100, -100)$
wife chooses Asia	$(-20, -20)$	$(100, 10)$

Table 5.4: Battle of the sexes.

Clearly the best for both is to choose the same vacation, either
to travel to Asia or visit ballparks. One of the couple will be
extremely happy with the choice, the other less so, but still
happy (above 0 in a utility indicates happiness). But if self-
interest motivates, the husband might say to the wife, "Look, our
best choices are to go together on vacation. But I am going to
insist on the ballpark vacation. It is better for you to join me, as

your utility is +10, rather than dig in and insist on traveling to Asia, where you'll be unhappier, with utility −20, without me." The wife replies, "Yes, but if I stick to going to Asia, you would be better off joining me, as your utility will jump from −20 to +10.

They are at an impasse! Based on the utilities, what will each do? The husband will likely stick to his guns and go to the ballparks, while his wife will not bend and will go to Asia, really the worst joint outcome—neither is happy! As they each travel off, they might each have some remorse, and seeing that they can at least improve their utility by unilaterally changing their minds, each decides to switch—the wife to the ballpark holiday, the husband to the Asia holiday. So the wife takes a flight to meet her husband at Yankee Stadium, only to reach him by phone as he simultaneously lands in Paris. Their utilities have fallen to the worst for both, −100. They are so unhappy— unhappier than they would be under any other joint choices. I hope they can at least laugh about it a few years down the road.

The Prisoners' Dilemma

OK, if the last two real-life decisions weren't enough to rattle you, let's consider the most slippery one, what is known as the **Prisoner's Dilemma**. This "paradox-in-a-game" was first raised in 1950 by two mathematicians, Merrill Flood and Melvin Dresher, who were working at the time for the RAND corporation, a policy and decision making think tank in the U.S.. Another mathematician, Albert Tucker, rephrased the problem and named it.

Here is how the situation goes. Imagine you and I are suspected by the District Attorney of committing a crime, but she doesn't have enough proof to convict us—she really wants a confession from both of us. We are taken into custody and here is what she offers:

- If neither one of us confesses, she will find some other lower-level charges and get us each sent to prison for 2 years.

- If we both confess, she will ensure we both get 10 years in prison.

- If you confess and I don't, then she will let you go free, but throw the book at me and have me sent away for 20 years.

- If I confess and you don't, the reverse will hold—she will let me go free, but have you sent away for 20 years.

We take as utilities the negative of the number of years we'll have to spend in prison (negative, as we want to avoid them)—see Table 5.5.

	I don't confess	I confess
you don't confess	(−2 years, −2 years)	(0 years, −20 years)
you confess	(−20 years, 0 years)	(−10 years, −10 years)

Table 5.5: Prisoners' Dilemma.

Now she leaves us. Perhaps surprisingly, it doesn't matter whether we talk between ourselves or decide on our own. What happens? We both realize that the best joint decision is to not confess—we'll get only two years in prison each. Not great, but it could be much worse. But as the seconds tick down to the DA reappearing and having to finalize our choices, I begin to think, "While it is best for both of us to not confess, and we have agreed to do so, at the last minute, it is much better for me to switch and confess. That way I'll go free—I really don't want to spend any time in jail. It would make me a little sad leaving you to sit in jail for 20 years while I escape jail time, but better you than me!"

And what do you think? Exactly the same thing. We smile at one another as the district attorney returns. Hell breaks loose as at the last moment we both confess, and it is now the DA who smiles. We both get the worse joint outcome—10 years in the slammer.

We knew what the issue was, we knew what the best joint choice was, and yet we both ended up with a very bad outcome. And therein lies the paradox—two rational people, you and I, ending up with the worst choice for us. Self-interest led us there!

All of the last three games—Chicken, the Battle of the Sexes and the Prisoner's Dilemma—can actually occur often in real-life situations, and it is good to recognize them when they appear so that you and your partner can seek a way out. In the next section we throw out our selfish desires in the hopes of coming to good group decisions.

Exercises

Classify each of the games in Exercises 5.2.1 to 5.2.6 as (a) two-person or multi-person, (b) with perfect information or without perfect information, (c) zero-sum or not, and (d) ending after a finite number of steps or not (and possibly playing forever). (If you are unfamiliar with any of the games, please search out the rules.)

Exercise 5.2.1. Dots and boxes
Exercise 5.2.2. Checkers
Exercise 5.2.3. Poker
Exercise 5.2.4. Soccer
Exercise 5.2.5. Nim
Exercise 5.2.6. Craps

Exercise 5.2.7. Generalized Nim is played where you decide ahead of time the maximum number k of coins you can take on a turn (for regular Nim, we had $k = 5$). You must take away at least one coin (but at most k) on your turn, and again the person who takes the last coin wins.
(a) Play the game with $k = 2$ and 33 coins. Who wins?
(b) Argue that the best strategy is to always leave your opponent with a multiple of $k + 1$ coins, if you can.
(c) If there are n coins to begin with, what is the winning strategy—who wins, the first player or the second? When and why?
Exercise 5.2.8. The game "Count to Thirty" is played as follows: someone start off and counts "one" or "one, two." The second player counts off either one or two more consecutive numbers, up from the last one said. So, for example, if the first player counts "one, two," the second player can count "three, four," and the first player can then respond with "five," and so on. The first player to reach thirty wins.
   Find the winning strategy.
Exercise 5.2.9. Can you relate the game in the previous exercise to generalized Nim (in Exercise 5.2.7)?

Exercise 5.2.10. We play a game with three players. Each player is given a coin, which he or she can either flip or just turn one desired side face up. If the coins all show the same side (that is, all three coins show heads or all three show tails), then the game is immediately played over. Otherwise, two of the coins match and one is different, and the person whose side up is different from the others is deemed the winner.

(a) It seems like the best thing to do is for each to randomly flip their coin. Explain why this gives each person a 1/3 chance of winning.

(b) Is there an advantage to turning your coin up first and showing everyone? What will likely happen in this scenario?

(c) If someone else has already shown the side up on their coin, what should you do? How does this better your chances of winning, compared to part (b)?

The game **Chomp** is a two-person game played with a rectangular array of dots (or coins). At a player's move, he or she yells "chomp," and picks a dot, removing it and all dots up to the right of it. For example, for the 3×5 board below,

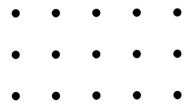

the first player's move could be to remove the dot in the second row, fourth column, and all dots that are <u>above and to the right</u> of this dot:

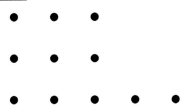

The following move, by the second player, could be to remove the dot in the first row, second column and all dots up and to the right of it.

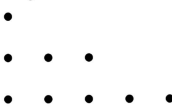

The game of Chomp ends when all the dots are gone, and the player who takes the last dot *loses.*

Exercise 5.2.11. Which player wins when you play with a board with just one row or just one column? What is the winning strategy?

Exercise 5.2.12. Find a winning strategy when you have a 2×5 board. Who wins?

Exercise 5.2.13. (Harder) Show that, provided the board has more than one dot, the first player has a winning strategy in Chomp. (Hint: One of the two players has a winning strategy. Argue that if it is the second player, then the first player could "steal" that strategy to use first!) [What the winning strategy is for the first player is unknown for most games of Chomp.]

Exercise 5.2.14. Suppose that we are playing dots and boxes (I am "J," you are "Y"). It is now your move. What is your best play?

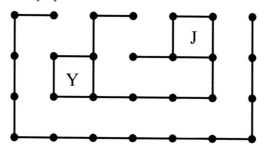

Exercise 5.2.15. Suppose you and a friend each slaps a quarter on the table, either heads or tails up (the player's choice). If the coins show the same side up, you win, but if they differ, your friend wins. Winner keeps both quarters. What do you think is the best strategy for both of you to play?

Exercise 5.2.16. Consider the following game matrix for a zero-sum game:

	R_1	R_2	R_3	R_4
S_1	2	3	-2	0
S_2	6	6	-1	4
S_3	-1	0	3	3

(a) Are there any dominated strategies?
(b) Are there any dominating strategies?
(c) What is the best strategy for each player?

Exercise 5.2.17. Consider the following game matrix for a zero-sum game:

	R_1	R_2	R_3	R_4
S_1	2	-3	-2	-3
S_2	1	-6	-1	-4
S_3	-1	0	-3	-3

(a) Are there any dominated strategies?

(b) Are there any dominating strategies?

(c) What is the best strategy for each player?

Exercise 5.2.18. Consider the following game matrix for a game (which is not zero-sum):

	R_1	R_2
S_1	$(60, 80)$	$(80, 50)$
S_2	$(10, 60)$	$(50, 10)$

What is the best strategy for each player?

Exercise 5.2.19. Consider the following game matrix for a game (which is not zero-sum):

	R_1	R_2
S_1	$(40, 80)$	$(80, 25)$
S_2	$(90, 40)$	$(25, 90)$

Are there any dominating strategies?

Exercise 5.2.20. Suppose we have two countries, A and B. Country B is a staging ground for terrorists who attack country A often. Country A has the option of invading country B to try to rout out the terrorists, or not. The terrorists in country B can attack country A or decide to hold a ceasefire. Here are some utilities for each decision:

	stage attacks on A	hold to ceasefire
invade B	$(-100, -100)$ years	$(400, 0)$
don't invade B	$(0, 400)$	$(100, 100)$

(a) What game is this like?

(b) What will each country do?

Exercise 5.2.21. Suppose two students, Sam and Clara, are accused of cheating on a final examination. The professor offers them the following choices, to admit to cheating or not to admit:

	Clara admits	Clara doesn't admit
Sam admits	both Sam and Clara receive a D in the course	Sam rewrites an open book exam, Clara receives an F
Sam doesn't admit	Sam receives an F, Clara rewrites an open book exam	both Sam and Clara rewrite a closed book exam

How will this play out?

Exercise 5.2.22. There are two candidates, Alice and Don, running for office. They can each choose to run negative campaigns or not. Here are some utilities for each:

	Don runs a negative campaign	Don doesn't
Alice runs a negative campaign	$(-500, -500)$	$(1000, -100)$
Alice doesn't	$(-100, 1000)$	$(200, 200)$

(a) Justify the relative utilities for each.

(b) What will happen in the election campaign? Why?

Exercise 5.2.23. Candidates Alice and Don, during the election campaign, decide to have a debate. There is one contentious issue, gun control, and they have a choice of supporting gun control, not supporting gun control, or evading the issue (which politicians are very good at!). Here are some utilities, given in terms of the percentage of the electorate that would vote for Alice (in an election between two candidates, it is a zero-sum game).

	Don supports gun control	Don doesn't	Don evades issue
Alice supports gun control	42%	5%	62%
Alice doesn't	48%	57%	52%
Alice evades issue	44%	6%	69%

What should each candidate do?

Exercise 5.2.24. Consider the following game matrix for a zero-sum game between players Tessa and Roseanne (Tessa has strategies T_1, T_2 and T_3, while Roseanne has strategies R_1, R_2 and R_3):

	R_1	R_2	R_3
T_1	-5	16	-50
T_2	-2	6	3
T_3	-3	-7	12

(a) Show that there are no dominated strategies.

(b) What is the minimum that Tessa can win/lose in each strategy? Which strategy is best in this regard?

(c) If Roseanne believes that Tessa will choose the strategy that maximizes the least she can win, what will Roseanne do?

(d) Show that if Tessa chooses strategy T_2 and Roseanne chooses strategy R_1, neither has any incentive to change their strategy if the other doesn't. (In this sense the strategies are **stable**.)

Exercise 5.2.25. Consider the following bidding game. One hundred dollars is put up for auction. You and I bid for the $100. The one who bids highest wins the $100, of course, but the loser has to pay his last bid, even though he doesn't get anything. For example, if you bid $25, then I bid $45, and you bid $55 dollars, you get the $100, and I have to pay $55, and I get nothing!

(a) Play the game with a friend a few times. What happens when you play? Is it surprising?

(b) Can you think of some reasonable scenarios in real life for this game?

Exercise 5.2.26. Imagine a multi-player game where everyone writes a positive number on a page; the person who wrote down the *smallest* number wins that many dollars.

(a) Play the game a few times. What happens?

(b) Is there a best strategy? If so, what is it?

Exercise 5.2.27. Two gas stations are on opposite corners in a town. Each one can see the price set by the other. If each wants to have the lowest price to attract the most customers, what will happen to the price of gasoline? In real life, what keeps this outcome from happening?

Exercise 5.2.28. When the Prisoner's Dilemma is played repeatedly, one can take on a strategy where you start by choosing to not confess, and if your opponent confesses, you punish your opponent in the next round by confessing, no matter what he or she does. Try out this strategy with a friend and see whether it leads to more cooperation over the long run.

Making Joint Decisions

We've seen how to take into account value and risk for our own individual decisions, and how to compete, but what about decisions we need to make in a group? That group may be a family, a sports team, a student council, a government body or a nation. The question remains—on what principles and with what process can we make joint decisions? Here we are interested in coming to some agreement as to what a *group*, not an *individual*, decides. We think of everyone having a vote, a choice to make individually. How we put together the votes forms the basis for this section.

If fairness is not the goal, there are a lot of ways to make decisions. One way is to have a **dictator**, someone who decides for everyone. This is the way many countries operate. One step away from this is to assign sets of votes to individuals, but not the same amount to each. This works out well if you want to give the illusion of democracy but really have something closer to a dictatorship. For example, suppose you and a friend are camp counselors in charge of a group of seven young children. Giving each child a single vote but giving yourself and your colleague each four votes may appear fair, and no one is a dictator, but in reality, you and your colleague together hold a majority of

0

the votes, and the children can't decide anything without at least one of you being in agreement.

When deciding between two options (which can be considered yes/no votes), a **coalition** is a collection of voters; the coalition is called a **winning coalition** if whenever they decide to vote for one option, the option passes (otherwise, the coalition is called **losing**). So in our camper example, the winning coalitions are exactly the groups that include both counselors, as they have 8 votes between them, out of 15 votes, which means they have a majority of votes, or contain a counselor and at least four campers (again they have at least 8 votes out of 15).

The assigning of a different number of votes to individuals may not seem fair, but it is often used in practice, and if you have been in charge of young children before, you can imagine the damage that true democracy can do in that situation.

Looking for Fairness

OK, certainly in general we want a voting system to be fair. But what does fair mean? Here are a few suggestions for cases where we are voting on just two candidates or alternatives, A and B:

1. **(anonymity)** We'd want our system to not depend on who votes a certain way, but only what way people vote—every voter should be treated equally.

2. **(neutrality)** We shouldn't treat one candidate differently from another—if we switched the votes for everyone, then if A was the winner previously, then B should win under the switch, and vice versa.

3. **(monotonicity)** Getting more votes can never be bad—if one candidate wins with some set of votes, then he or she wins with more voters selecting him or her.

These conditions seem pretty universal, don't they? Certainly, if you have an odd number of voters, then in any vote by everyone, one candidate must get a majority of the vote (this may or may not happen if there is an even number of voters, as the vote can be split right down the middle between the two candidates).

The **majority rule**, that the candidate with the most votes wins, satisfies all three of these properties. There is a famous result by mathematician Kenneth May that says that *any* group decision by an odd number of people that satisfies the three properties and always declares a unique winner must be majority rule—there are no other choices to follow.

The argument actually isn't so hard. Consider the smallest group of voters that ensures that A wins; suppose this group has size n. By property (1) any group of n voters who vote for A ensures that A wins, and any smaller group ensures that B wins. Furthermore, property (3) ensures that if A gets more than n votes, A wins, and if A gets less than n votes, A loses to B. From property (2) the same holds for B. Thus the group decision process is what is called a **quota system**—all you need to win is to have a certain number of votes (this certain number is called the **quota**). Here, the quota is n. Finally, if there is an odd number P of people, we see that n must be $(P+1)/2$, because (i) if n were smaller than $(P+1)/2$, then n is smaller than $P/2$—half the total—as P is odd, and it is possible for both A and B to meet the quota and therefore both be winners (which isn't allowed), while (ii) if n were greater than $(P+1)/2$, then it would be possible for neither candidate to meet the quota, and therefore there be no winner, again something that is not allowed. It follows that n is $(P+1)/2$, a fraction more than half, and what we have is majority rule.

As an example, if $P = 101$, that is, there are 101 voters, then $(P+1)/2 = 102/2 = 51$, the next integer up from half of the total. If the quota were smaller than 51, then it is possible for both A and B to win, when one candidate gets 50 and the other gets 51 votes. If the quota were greater than 51, then again with one candidate getting 50 and the other getting 51 votes, we'd have a problem, but then neither would win. Thus the quota must be 51—a majority of the votes.

So what happens if there are an *even* number of voters? In that case we are stuck with no process. The argument we've given says that any possible system must be a quota system, but we are left with no good quota to use—half or less allows both A and B to be winners, while more than half allows for no winners. This is our first example of what we will see throughout this section—sometimes there is *no* good way to decide on a winner.

Plurality—When More Is Just Enough

Now we turn to having more than two candidates, and things get murkier quickly. There need not be a winner under majority rule—not only might no candidate get a majority, but each candidate might be far from a majority. For example, if out of

100 000 voters voting for candidates A, B and C, what happens if
A gets 33 657 votes, B gets 34 082 votes and C gets 32 261 votes?
No one has received a majority (50 001) votes, or even close to it.
What is often applied is the **plurality rule**—the candidate with
the most votes wins (if there is a tie with the most votes, all the
top vote getters are declared winners, and some tie-breaking
process should be used). But there is something unappealing
about the plurality rule—in the case I just mentioned, we would
declare B the winner with the most votes, 34 082, but most
people—almost 66%—didn't vote for B at all!

The Plurality System: The candidate with the most first place
votes is declared the winner (when there is a tie at the top, a
suitable adjustment must be made).

You might think that the previous example was artificial and
wouldn't occur in real life, but during the 2000 presidential
elections in the U.S., the outcome in the deciding state, Florida,
was as in Table 5.6.

Candidate	Total vote
Bush, George W. (Republican)	2 912 790
Gore, Al (Democrat)	2 912 253
Nader, Ralph (Green)	97 488
Buchanan, Patrick J. (Reform)	17 484
Browne, Harry (Libertarian)	16 415
Hagelin, John (Natural Law/Reform)	2281
Phillips, Howard (Constitution)	1371
Other	3028

Table 5.6: Florida voting outcome in the federal election.

The total number of votes in Florida was 5 963 110. Each state
has a "winner takes all" approach, with the plurality winner,
in general, taking all of the state's electoral votes (the electoral
votes, in the Electoral College, decide on the presidency). George
W. Bush won the hotly contested state, and with Florida's 25
electoral votes, received 271 electoral votes, exactly one more
than the minimum needed to be elected president (the Electoral

College has 538 seats, so anything more than $538/2 = 269$ is a majority).

Across the U.S., Gore actually won the popular (that is, overall) vote: Gore received 50 999 897 votes, more than half a million more than Bush's 50 456 002. More strikingly, it is widely felt that had Ralph Nader not been on the Florida ballot, his supporters would largely have moved to Gore and changed the winner of the presidential election. From a Democratic point of view, sometimes too many candidates spoil the voting broth!

The Borda Count—Taking into Account Rankings

To have the winner reflect more of the will of the people, what is often done is to ask voters not just to vote for a winner, but rank the candidates, from lowest to highest. For example, one voter might rank (that is, order) the three candidates A, B and C as

$$B \prec A \prec C,$$

meaning that the voter prefers A to B and C to A (and of course, C to B); this listing is called the **voter's profile**. A vote consists now of a list of voter profiles for all voters. With three candidates, there are six possible voter profiles for the candidates:

$$A \prec B \prec C, \quad A \prec C \prec B, \quad B \prec A \prec C,$$
$$B \prec C \prec A, \quad C \prec A \prec B, \quad C \prec B \prec A.$$

The multiplicative counting principle gives this as well, as we are looking for arranging three objects in order, and this can be done in $P_{3,3} = 3! = 6$ ways. For 10 candidates, there are $10! = 3\,628\,800$ many possible rankings of the candidates, so the number of possible rankings grows huge quite quickly. With lots of voters, one often combines the different orderings into a table, listing how many voters had the candidates in a particular order. Table 5.7 shows an example (the rank numbers start at the top with 1 and head downwards); there are only six different voter profiles, as many people agreed on the six rankings.

The **Borda count system** chooses a winner from a set of voter preferences by attaching a weight to each rank value: rank 1 out of n candidates gets the highest weight, $n - 1$, rank 2 gets the next

| | # of voters with this profile | | | | | |
rank	8	5	13	9	4	8
1	A	B	C	C	E	E
2	C	A	A	B	D	B
3	B	C	E	E	A	A
4	E	E	D	A	C	D
5	D	D	B	D	B	C

Table 5.7: Table of voter profiles from a vote.

highest weight, $n - 2$, and so on, down to the lowest rank, which gets a weight of 0.

| | # of voters with this profile | | | | | | |
rank	8	5	13	9	4	8	Borda weight
1	A	B	C	C	E	E	4
2	C	A	A	B	D	B	3
3	B	C	E	E	A	A	2
4	E	E	D	A	C	D	1
5	D	D	B	D	B	C	0

Table 5.8: Table of a vote with Borda weights.

I put the Borda weights in the right column; they run in reverse order to the ranks. Now the Borda count of a candidate is the sum of all of the weights for the candidate over all of the voting profiles. So in our example, for candidate A, there are eight profiles where A has rank 1, and hence Borda weight 4, so these add $8 \times 4 = 32$ to A's count. There are $5 + 13 = 18$ profiles where A has weight 3, $4 + 8 = 12$ where A has weight 2, 9 where A has weight 1, and 0 where A has weight 0. Thus A's Borda count is

$$8 \times 4 + 18 \times 3 + 12 \times 2 + 13 \times 1 + 0 \times 0 = 119.$$

An easy way to get this is for each location of the candidate in the table, multiply the number at the top of the column (the number of votes of the candidate in this profile) by the number at the right (the Borda weight of the position) and add them up:

candidate A's Borda weight

$$= 8 \times 4 + 5 \times 3 + 13 \times 3 + 9 \times 1 + 4 \times 2 + 8 \times 2$$
$$= 119.$$

Similarly, *B, C, D, E* have Borda counts 87, 126, 24, 105, respectively. Under the Borda count system, the winner is the candidate with the highest Borda count, which here is candidate *C*. The Borda count takes into account not only each voter's first choice for a winner, but the rankings over all. As well, we not only can declare a winner (or winners, if there is a tie at the top Borda count), but also a group ranking of candidates, by their decreasing order of Borda counts, with any tied candidates occupying the same position. So in our example, our group ranking, by Borda counts, is C, A, E, B, D.

The Borda count system was developed in 1770 by the French mathematician Jean-Charles de Borda. It is used in a variety of settings, such as electing members of Parliament of Nauru, an island in the South Pacific, as well as in sports, such as choosing Major League Baseball's Most Valuable Player (MVP) award and U.S. college football's Heisman Trophy.

The Borda Count System: From the voter profiles, calculate the Borda count for each candidate, summing the weights of the position of the candidate under each profile. The candidate with the highest Borda count is declared the winner (with a suitable adjustment when more than one candidate has the highest Borda count).

As appealing as Borda counts are, they can also have their problems. For example, consider the election results in Table 5.9 between A, B and C, along with their Borda weights:

	# of profiles			
rank	11	5	5	Borda weight
1	A	B	C	2
2	B	C	B	1
3	C	A	A	0

Table 5.9: Problem with Borda counts.

The Borda counts for A, B and C are 22, 26 and 15, respectively, so we should declare B the winner. But A was on top of more than half the votes, so shouldn't A win by the majority rule? What we learn is while the Borda count sounds nice, in practice it can lead to some unusual outcomes.

Condorcet Winners—Beating Everyone!

If we look back at Table 5.7,

rank	8	5	13	9	4	8
1	A	B	C	C	E	E
2	C	A	A	B	D	B
3	B	C	E	E	A	A
4	E	E	D	A	C	D
5	D	D	B	D	B	C

# of voters with this profile

we see something interesting: even though B won by the Borda count system, if we compare candidate A to each other candidate, seeing how many people voted for A over the other, we see that A always came out on top:

- A versus B: A is above B in $8 + 13 + 4 = 25$ profiles while B is above A in $5 + 9 + 8 = 22$

- A versus C: A is above C in $8 + 5 + 4 + 8 = 25$ profiles while C is above A in $13 + 9 = 22$

- A versus D: A is above D in $8 + 5 + 13 + 9 + 8 = 43$ profiles while D is above A in 4

- A versus E: A is above E in $8 + 5 + 13 = 26$ profiles while E is above A in $9 + 4 + 8 = 21$

So it seems to make sense to declare A the winner! Such a candidate is given a special title.

A Condorcet Winner: If, from the voter profiles, a candidate A would win in head-to-head contests with each of the other candidates, that is, for any other candidate X, a majority of voters rank A over X, then A is called a **Condorcet winner**.

Condorcet winners, if they exist in an election, would seem to be a good choice for a winning candidate. If someone wins a majority of first place votes, that person would certainly be a Condorcet winner, which is what we might hope for in a reasonable voting system. But sadly, there aren't always Condorcet winners. The simple set of profiles in Table 5.10 shows why:

The Condorcet winner was introduced by Nicolas de Condorcet (1743–1794), a French mathematician who was also a political scientist.

	# of voters		
rank	1	1	1
1	A	B	C
2	B	C	A
3	C	A	B

Table 5.10: List of profiles with no Condorcet winner.

Here A wins in a head-to-head contest with B, but B wins in a head-to-head contest with C and C wins in a head-to-head contest with A—all by 2 to 1 margins. So there is no one who wins in a head-to-head contest with everyone else, that is, there is no Condorcet winner. Our shining, brand new voting system, while it would seem to work well when it does work, goes down in flames even for a small example.

Running with Instant Run-Offs

We'll discuss one more proposal for finding a winner from voter's profiles, by using what is often called the **instant run-off system**. Under this process, one again takes the voter profiles, but proceeds differently from the Borda counting system. First one looks at the profiles and sees if there is a candidate who has won more than half of the first place votes—if so, that candidate is the winner and the process stops. Otherwise, the candidate/candidates with the lowest number of first place votes is/are removed from each profile, and the remaining candidates are moved up in position on each profile to fill any gaps. The process is repeated with the **reduced voter profiles** until a winner is declared.

So, for example, let's take the voter profiles from Table 5.7 again:

The Instant Run-off System is used in elections for the Australian House of Representatives and to select the presidents of India and Ireland, as well as for many other political elections in countries around the world.

rank	# of voters with this profile					
	8	5	13	9	4	8
1	A	B	C	C	E	E
2	C	A	A	B	D	B
3	B	C	E	E	A	A
4	E	E	D	A	C	D
5	D	D	B	D	B	C

By counting in the first row of the table, we see that A has 8 first place votes, B has 5, C has $13 + 9 = 22$, D has 0 and E has $+8 = 12$. Clearly D has the least number of first place votes, so D is removed from each voter's profile and the candidates below D are moved up one spot in each profile:

rank	# of voters with this profile					
	8	5	13	9	4	8
1	A	B	C	C	E	E
2	C	A	A	B	A	B
3	B	C	E	E	C	A
4	E	E	B	A	B	C

Now for these reduced voter profiles, we again look for a candidate with a majority of the first place votes. No such candidate exists (as the removed candidate D had no first place votes, no one moved up into the first spot, so the number of first place votes for each candidate remains the same as before). Now B has the fewest first place votes (5), so we remove B from each profile, forming new reduced voter profiles by moving up candidates below B on each profile by one spot:

rank	# of voters with this profile					
	8	5	13	9	4	8
1	A	A	C	C	E	E
2	C	C	A	E	A	A
3	E	E	E	A	C	C

Here A has 13 first places votes, C has 22 and E has 12, and so E leaves the profiles:

	# of voters with this profile					
rank	8	5	13	9	4	8
1	A	A	C	C	A	A
2	C	C	A	A	C	C

A now has $8 + 5 + 4 + 8 = 25$ first places votes compared to C's $13 + 9 = 22$. A, with the majority of first place votes on the reduced profiles, is declared the winner. This is markedly different from what happened under the Borda count (or plurality), where C was declared the winner.

The Instant Run-Off System: From the voter profiles:

- If there is a candidate who has won a majority of the first place votes, that candidate is declared the winner.

- Otherwise, the candidates with the lowest number of first place votes are removed from each profile, and the candidates below the removed candidates are moved up one position on each profile, to form the reduced voter profiles.

The process is repeated until a winner is declared.

So all of our proposed systems—plurality, Borda counts, Condorcet winners and instant run-off—all suffer from some serious problems when we have at least three candidates. Perhaps there is lurking, somewhere out there, a better system, one that will meet some very basic requirements that we feel any voting system should have?

An Arrow to the Heart of Voting

Economist Kenneth Arrow was interested in group voting that led to a group profile (that is, preference list) from individual voter profiles. He proposed in 1951 a few such rules for a voting system that seemed obvious and even essential for fairness:

1. **(universality)** Other than the fact that everyone was to rank the candidates, there were no other restrictions on how this was to be done by each voter (although the rankings were to be **transitive**, as described earlier: if voter X preferred A to B (that is, ranked A higher than B) and B to C, then she preferred A to C).

2. **(unanimity)** If *every* voter prefers candidate A to candidate B, then the voting group should prefer A to B.

3. **(non-dictatorship)** There shouldn't be a person X whose preferences completely determine the group's preferences. That is, it shouldn't be the case that for some person X, whenever X prefers candidate A to candidate B, so does the group.

4. **(independence of irrelevant alternatives)** The group's preferences between two candidates A and B should only depend on each of the voters' preferences between A and B, and not on individuals' preferences involving other candidates.

Perhaps the only one that is unclear is the last one, the independence of irrelevant alternatives. What this rule says is that when trying to decide whether the group should rank A higher than B or not, all that should matter is how each person ranks A compared to B; it shouldn't make any difference how individuals rank A with respect to some other candidate C, how B is ranked by voters with respect to C, or even how C ranks compared to other candidates different from A and B.

These four principles seem natural, desirable and even required if we are to form a voting system for group rankings from individuals' rankings, right? What Kenneth Arrow *proved* was that it was *impossible* to find any voting system that met rules 1-4! It's not just that we can't find any such voting system—no one can, and no one will ever be able to, if we need these few rules to be satisfied. So once again, we see that there are intrinsic problems with trying to decide how groups should vote based on the voting of each person.

Exercises

Exercise 5.3.1. Suppose you decide to give each of seven campers five votes each. How many votes should you give yourself and your friend (the two counselors) so that no decision can be made by the campers alone?

Exercise 5.3.2. Suppose now that you decide to give each of 13 campers two votes each. There are three counselors. How many votes should you give each counselor so that no decision can be made without the agreement of at least one counselor?

Exercise 5.3.3. Suppose now that you decide to give each of 13 campers two votes each. There are three counselors. How many votes should you give each counselor so that no decision can be made without the agreement of at least two counselors?

Exercise 5.3.4. Revisit the section's example where you and a friend are camp counselors in charge of a group of seven young children and you give each child a single vote but give yourself and your colleague each four votes. If any yes/no vote is decided by a majority of the votes (that is, at least 8 out of the 15 votes), describe the winning coalitions.

A **minimal winning coalition** is a winning coalition such that if *any* voter is removed from the coalition, the coalition is no longer winning. Every winning coalition contains a minimal winning coalition, so it is enough to just list the winning coalitions. Answer Exercises 5.3.5 through 5.3.7.

Exercise 5.3.5. What are the minimal winning coalitions for Exercise 5.3.1?

Exercise 5.3.6. What are the minimal winning coalitions for Exercise 5.3.2?

Exercise 5.3.7. What are the minimal winning coalitions for Exercise 5.3.3?

Exercise 5.3.8. Rather than just require a majority of the votes, one might require more than that, a quota, as described earlier. For example, in 1958 the Treaty of Rome, signed by France, Germany, Italy, Belgium, The Netherlands and Luxembourg, set up a voting system, called the European Economic Community, for yes/no questions affecting the countries. Votes were assigned as follows:

Country	# votes
France	4
Germany	4
Italy	4
Belgium	2
The Netherlands	2
Luxembourg	1

In order for an item to pass, it needed to receive at least 12 of the 17 possible votes. What are the minimal winning coalitions?

Exercise 5.3.9. Is the voting system described in the previous exercise a quota system?

Exercise 5.3.10. There are 435 members in the House of Representatives and 100 members of the Senate. The president and vice-president are also voting members, giving a total of 537 voters for yes/no bills. A bill passes through the House or the Senate with a majority vote, and needs to pass through both. The vice-president votes in the Senate just in case there is a tie there, and the president has veto power over bills that can be overridden if the vote is more than two-thirds in each of the House and Senate, that is, supported by at least 67 Senators.

Show that the minimal coalitions are

1. 218 representatives and 51 senators and the president

2. 218 representatives, 50 senators, the vice-president and the president

3. 290 representatives and 67 senators

Exercise 5.3.11. Is the U.S. system, described in the previous exercise, a quota system?

Exercise 5.3.12. In the 1998 Minnesota gubernatorial election for governor, the results were as follows:

candidate	# votes
Norm Coleman	717 350
Thomas Fiske	787
Frank Germann	1932
Hubert H. Humphrey III	587 528
Fancy Ray McCloney	919
Ken Pentel	7034
Jesse Ventura	773 713
Chris Wright	1727

(There were also 776 write-ins as well.)

(a) What percentage of the vote did each candidate get?

(b) The winner was decided by plurality. Who won?

(c) What problem do you see here with using the plurality system?

Exercise 5.3.13. If there are five candidates in an election, how many possible voter profiles are there?

Exercise 5.3.14. Consider the following voter profiles in an election with candidates Alisha, Ben and Charlotte:

	# of profiles					
rank	6	7	5	2	1	3
1	A	B	C	A	B	C
2	B	C	B	C	A	A
3	C	A	A	B	C	B

The winner is to be decided by the Borda count.

(a) Are there any other possible voter profiles from the six listed? Why?

(b) Who wins under the plurality system?

(c) Calculate the Borda count for each candidate.

(d) Who wins under the Borda count?

Exercise 5.3.15. Consider the following voter profiles in an election with candidates Alice, Bob, Cynthia, Don and Ed:

	# of voters with this profile					
rank	4	15	8	6	3	10
1	A	B	C	C	E	E
2	C	A	A	B	D	B
3	B	C	E	E	A	A
4	E	E	D	A	C	D
5	D	D	B	D	B	C

The winner is to be decided by the Borda count.

(a) Calculate the Borda count for each candidate.

(b) Who wins?

Exercise 5.3.16. Consider the following election profiles for candidates Mateo, Keisha and Sofia:

	# of profiles			
rank	712	855	928	303
1	M	K	S	M
2	K	S	K	S
3	S	M	M	K

(a) Calculate the Borda count for each candidate.

(b) Who wins under the Borda count system?

(c) Who wins under the plurality system?

Exercise 5.3.17. Explain why, if a candidate wins a majority of first place votes, he or she is a Condorcet winner.

Exercise 5.3.18. Can there be more than one Condorcet winner? Explain your answer.

Exercise 5.3.19. For the voter profiles in Exercise 5.3.15, is there a Condorcet winner?

Exercise 5.3.20. For the voter profiles in Exercise 5.3.15, who wins under the instant run-off system?

Exercise 5.3.21. For the voter profiles in Exercise 5.3.16, who wins under the instant run-off system? Why?

Exercise 5.3.22. What difficulty do you run into with the voter profiles in Exercise 5.3.14 if you want to use the instant run-off system?

Exercise 5.3.23. Consider the following election results between political parties X, Y and Z:

	# of profiles		
rank	42	38	20
1	X	Z	Y
2	Y	X	Z
3	Z	Y	X

(a) Who wins under the instant run-off system?

(b) Who wins under the plurality system?

(c) Who wins under the Borda count?

Exercise 5.3.24. Which of universality, unanimity and independence of irrelevant alternatives does a dictatorship satisfy?

Exercise 5.3.25. Explain why the Borda count system satisfies unanimity.

Exercise 5.3.26. Does the Borda count system satisfy the independence of irrelevant alternatives? Explain your answer.

Exercise 5.3.27. (Harder) Suppose in Exercise 5.3.15 that the number of voters with the first profile is not 4, but not known precisely until after a recount; all of the others are exact.

(a) Show that no matter what, the winner under the Borda count is either A or B.

(b) Under what conditions (that is for what number of voters with the first profile) does A win?

And Now the Rest of the Story …

*The reason that countries have such a tough time agreeing or sticking to a global warming protocol is much more than political will—it is a question of game theory. Global warming is one big multi-player version of the Prisoner's Dilemma, but with huge stakes. Of course, it is in all countries' best interest to decrease emissions, but it gets dicier if one country can get away with **not** decreasing emissions while everyone else is. Simplifying the situation for the sake of discussion, imagine there are only two countries, A and B, which each has to decide whether to decrease greenhouse gas emissions. Here is a game matrix, with some reasonable utilities:*

	B reduces emissions	B doesn't reduce emissions
A reduces emissions	$(1000, 1000)$	$(-10\,000, 10\,000)$
A doesn't reduce emissions	$(10\,000, -10\,000)$	$(-5000, -5000)$

The actual utilities assigned don't make a difference, but the important thing is that while the best joint decision is for both to reduce emissions,

and the worst joint decision is for neither to, when a country believes
that another will reduce its emissions, it is in its best interest to not
reduce emissions (for then it can reap economic benefits compared to a
country that is spending resources on reduction). In this version of the
Prisoner's Dilemma, whether for just two countries or many, there is an
enormous pull to not agree to reduce, just based on self-interest.

So what can be done? Does game theory say that we are doomed to
never get agreement? One way out may be to change the utilities for
each country, or really, for each country's leadership. Voters can let
candidates know that they value adhering to greenhouse gas reductions,
even if they are the only ones. Their utility may go down in such a case,
but not as drastically as before. Perhaps citizens around the world could
revise the game matrix for global warming:

	B reduces emissions	B doesn't reduce emissions
A reduces emissions	$(1000, 1000)$	$(-100, 100)$
A doesn't reduce emissions	$(10\,000, -10\,000)$	$(-5000, -5000)$

For this game matrix, there is no incentive for either country to shift
from reducing emissions, as any unilateral shift will decrease their
utility.

While decision theory and game theory may be the essence of the problem
for global warming, who is to say that they may not also be the solution
as well?

Review Exercises

Exercise 5.4.1. What is the expected number of pips shown when two dice are rolled?

Exercise 5.4.2. What is the expected number of pips shown when three dice are rolled?

Exercise 5.4.3. Suppose that you plan to invest $5000 in a stock for a year. From previous
behavior of the stock, you think that

- with probability 0.10 the stock will increase by 20%,

- with probability 0.60 the stock will increase by 5%, and

- with probability 0.30 the stock will lose 15% of its value.

(a) What can you expect to earn (or lose) if you do invest for the year?
(b) Is it worthwhile to invest?

(c) What would be the most you would be willing to pay a stock broker each year to invest your money repeatedly, if you expect to earn, on average, 3% on your investments?

Exercise 5.4.4. Suppose a person has assigned a utility of 200 to $5000. If she is indifferent between getting $5000 outright or a lottery where she can win $10 000 with probability 0.25 and nothing with probability 0.75, what utility should she assign to $10 000?

Exercise 5.4.5. Suppose you have a choice between a trip to London or a trip to Rome. If you go to London, there is a 40% chance you'll overspend your budget for the trip; if you go to Rome, the chance that you'll overspend is only 20%. Now if you go to London and overspend, there is a 75% chance you'll have a lot of fun, but if you go to London and don't overspend, there is only a 45% chance of having fun. If you go to Rome and overspend, you estimate your chance for fun is 65%, while if you go to Rome and don't overspend, there is a 55% chance of having fun. You attach utilities as follows:

outcome	utility
go to London, overspend, have fun	500
go to London, overspend, don't have fun	−200
go to London, don't overspend, have fun	1000
go to London, don't overspend, don't have fun	100
go to Rome, overspend, have fun	800
go to Rome, overspend, don't have fun	−300
go to Rome, don't overspend, have fun	1500
go to Rome, don't overspend, don't have fun	−100

(a) Are the overspending and fun independent of one another, once a location for the trip has been chosen?

(b) Draw a decision tree.

(c) Find the maximum expected utility (MEU). What decision should you make based on MEU?

Exercise 5.4.6. Suppose you have a choice between receiving $10 000, a lottery where with probability 0.50 you get $30 000 and get nothing otherwise, and a third lottery where you get $100 000 with probability 0.10 and $1000 with probability 0.90. Suppose also that your utilities for an amount of money are equal to its value.

(a) If you use MEU, which do you choose?

(b) If you are a satisficer where $5000 is enough for you, which do you choose?

Exercise 5.4.7. Consider the following game matrix for a zero-sum game:

	R_1	R_2	R_3
S_1	2	-3	-2
S_2	1	-6	-1
S_3	-1	0	-3
S_4	4	1	1

(a) Are there any dominated strategies?

(b) Are there any dominating strategies?

(c) What is the best strategy for each player?

Exercise 5.4.8. Consider the following game matrix for a game (which is not zero-sum):

	R_1	R_2
S_1	$(3,3)$	$(1,4)$
S_2	$(4,1)$	$(2,2)$

(a) What is the best joint strategy, if they decide to maximize the sum of their utilities?

(b) Knowing this, what does each player do, if they have to jointly choose, with no option to change their decision?

(c) If they can, after seeing their opponent's choices, change their own choice as many times as they like, what will happen in the game?

Exercise 5.4.9. The United Nations' Security Council consists of five permanent members—U.S., France, England, China and Russia—together with ten other non-permanent members (the latter change over the years, serving two-year terms). A resolution passes only when at least 9 of the 15 members vote in favor, but each of the 5 permanent members has a veto (and can keep the resolution from passing).

(a) Describe the minimal winning coalitions.

(b) Suppose that we give each non-permanent member one vote, and each permanent member seven votes. If we set the quota to be 39 for the passing of a resolution, show that this system matches the voting rules for the Security Council.

Exercise 5.4.10. Consider the following voter profiles in an election with candidates Alice, Bob, Cynthia, Don and Ed:

	# of voters with this profile					
rank	23	12	14	8	13	18
1	A	B	C	D	D	E
2	B	D	A	B	E	B
3	C	C	E	E	A	A
4	D	E	B	A	C	C
5	E	A	D	C	B	D

(a) Does anyone have a majority of first place votes?

(b) Who wins under the plurality system?

(c) Who wins under the Borda count system?

(d) Is there a Condorcet winner?

(e) Who wins under instant run-off?

Art Imitating Math

Maurits Cornelis, known as "Mauk" or just "MC", was born in 1898 in Leewarden, The Netherlands. From an early age, little Mauk was obsessed with the idea of filling up space with non-overlapping objects. Indeed, when he was small, he would, while making a sandwich, choose very carefully the shape and sizes of the cheese so that the pieces would fit precisely together to cover the bread. A very interesting preoccupation, indeed!

At school, Mauk didn't excel and didn't do particularly well in mathematics and science, though he did like the preciseness and methodologies of the subjects—after all, his father and three of his brothers were in engineering fields, and he appreciated logical lines of thought.

On the other hand, Mauk loved art at school, and in school learned how to make linoleum cuts for printing, moving into wood cutting soon afterwards. Though he had plans to study architecture at the School for Architecture and Decorative Arts in nearby Haarlem, one of the instructors there, Samuel Jessurun de Mesquita, recognized Mauk's artistic promise and took him under his wing to study as a graphics artist.

MC may not have done well in school, but he was always a student—reading, exploring, learning, and especially carrying out artwork exercises that he assigned himself. Some were based on his travels. He loved Italy and the artwork there and was even more drawn to Granada, Spain, where he was fascinated by the Alhambra, a fourteenth century palace, filled with geometric artwork, polygons that filled his early fixation with space filling.

MC loved the art in the Alhambra, but he noticed that it was rather abstract—there weren't any animals to be found and hence lacked some "life" to it. He furiously began days, months and years of sketches, trying to reproduce the geometry he loved with life forms playing a central role.

He started creating artwork based on the sketches, each one with a new twist. He loved the work, the self-learning, even though it was a rather *lonely time—no one was taking the same approach, mixing geometry and life. His studies in those sketchbooks were personal development— he could look back and see how his understanding of the underlying principles had grown. But he felt there was still more, beyond his grasp. Then, in 1937, during a trip to visit his parents, he showed his half brother, nicknamed "Beer," some of his latest artwork. Beer was a Professor of Geology at the University of Leiden and immediately saw a connection to what he called "crystallography." Beer wrote his brother a few days later that he had come across a mathematical research paper by someone named George Pólya, on designs where a small figure is repeated over and over again, in some regular way, to cover the plane. The paper went on to describe the 17 ways this could be done. The details would be beyond Mauk, but he found much to comprehend in the accompanying figures in the article. In an instant, a whole new world was opening up to MC, and he was ready to walk through that door.*

WHAT COULD POSSIBLY BE MORE OPPOSITE TO MATHEMATICS THAN ART? Art is emotional, intuitive and creative, while mathematics seems logical, difficult and often done by memorization. But this is a false view of mathematics. Indeed, if mathematics is understood properly and done right, it is emotional, intuitive and creative as well. Both art and mathematics center on patterns—art seeks to develop and integrate them, while mathematics aims to comprehend and analyze them. And the same kind of work required to excel at mathematics is necessary to produce the highest quality artwork. So even if you plan to be an artist, there is much to learn from mathematics.

6.1 *Math That Makes the Art*

What is beautiful? Artists have been preoccupied with this question ever since they began painting on cave walls. While beauty is in the eye of the beholder, it is also true that there are some things that people in general find attractive.

> "Without mathematics there is no art."—Luca Pacioli (1445–1517), an Italian mathematician who collaborated with Leonardo da Vinci.

Mona Lisa Is not so Square

We like to look at objects that are in good proportion—a building, a flower, a person. But what is good proportion? Certainly an object that is too wide or too skinny is not as nice to look at as one that is *just right*. There have been studies as to what kind of rectangles we find attractive. Back in 1876, psychologist Gunter Fechner surveyed a large group of people to have them compare and state a preference among some common shaped rectangles (see Figure 6.1). Figure 6.3 shows the results.

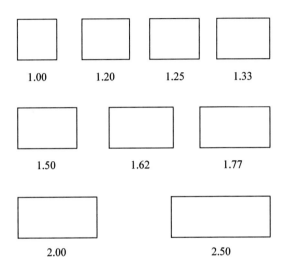

Figure 6.1: Rectangle shapes for G. Fencher's preference experiment.

1.00   1.20   1.25   1.33

1.50   1.62   1.77

2.00   2.50

So the middle rectangle in Figure 6.1 is by the far the preferred one. But the ratio of the length to width is an unusual number, 1.62. All of the others arise as small number ratios—1.00 is 1 : 1, 1.20 is 6 : 5, 1.25 is 5 : 4, 1.33 is 4 : 3, 1.5 is 3 : 2, 1.77 is 23 : 12 (not so small, but there is a choice nonetheless), 2.00 is 2 : 1, and 2.50 is 5 : 2. So why the preference for 1.62, and where does it come from?

The ancient Greeks knew. Numbers to them were constructions, and they built a nice rectangle in the following way. They started with a square $ABCD$, each side of length 1, and cut it in half, as in Figure 6.4, with new points E and F. They then put a compass at F and drew a circle with radius equal to the distance from F to one of the opposite corners, say B. This circle cuts the line extending the side CD at a point which we'll call X. We'll extend the opposite side, AB, so that a vertical line from X meets it, at point Y (see Figure 6.4). Then $AYXD$ is an example of a golden rectangle.

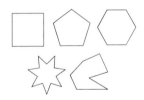

Figure 6.2: Rectangles are examples of **polygons**, which are two-dimensional figures made by joining points in succession with straight lines, returning to the starting point. Other examples include parallelograms, pentagons, hexagons and stars. A polygon is **regular** if all sides are equal in length and all angles are equal (so a square is a regular, as is an equilateral triangle); the polygons in the top row are all regular, while the ones in the bottom row aren't.

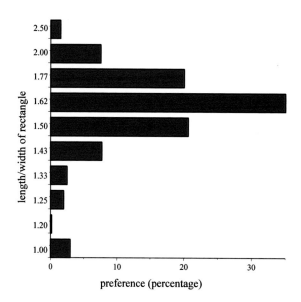

Figure 6.3: Data from G. Fencher's preference experiment.

The golden rectangle has many nice properties. The ratio of its width to its height is called the **golden ratio** and is often written by the Greek letter ϕ (spoken as "FIY"). With a bit of geometry you can show that it is the number $\dfrac{1+\sqrt{5}}{2}$, that is, $\dfrac{1}{2} + \dfrac{\sqrt{5}}{2}$ (see the exercises), which is not a rational number. It is approximately 1.618, or 1.62 to two decimal places, and we see why Fencher had this ratio among his options. Moreover, if you look at Figure 6.4, you find more than one golden rectangle—$BYXC$ also turns out to be a golden rectangle as well!

The number ϕ has many fascinating properties. For example, the square of the golden ratio is one more than the golden ratio, that is,

$$\phi^2 = \phi + 1,$$

as

$$
\begin{aligned}
\phi^2 &= \left(\frac{1+\sqrt{5}}{2}\right)^2 \\
&= \frac{(1+\sqrt{5})(1+\sqrt{5})}{4}
\end{aligned}
$$

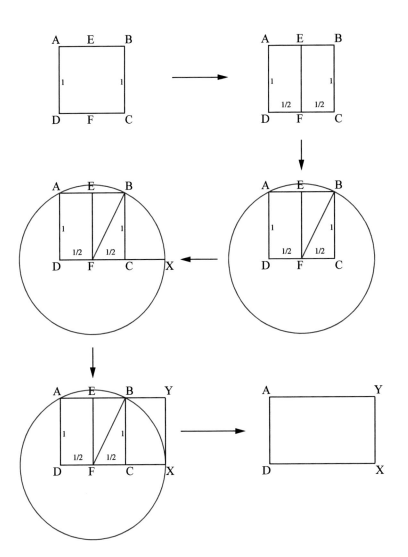

Figure 6.4: Building the golden rectangle.

trilliums

geraniums

Figure 6.5: Some flowers and their Fibonacci petals.

mountain avens

black-eyed Susans

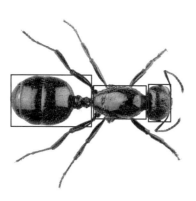

Figure 6.6: Black garden ant and golden rectangles.

The properties of the golden ratio were very pleasing to the Greek mathematicians who discovered them, and the golden rectangle held a very special place in their hearts, as the ideally shaped rectangle. This love of the golden rectangle extended to many Renaissance painters, especially Leonardo da Vinci. Da Vinci was not only a prolific artist, but was a scientist and engineer as well, filling many notebooks with his ideas and sketches—he was indeed a student of the arts and sciences. His most famous painting is undoubtedly the **Mona Lisa**, and even here we can find instances of the golden rectangle lurking about (see Figure 6.7), as the golden ratio is naturally present as an ideal proportion in the human body.

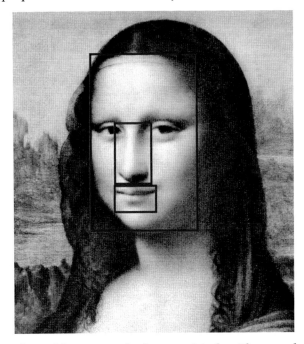

Figure 6.7: Leonardo da Vinci's *Mona Lisa*, with golden rectangles superimposed.

Love for golden rectangles has persisted, with many buildings incorporating the proportion in various ways in their construction. The Parthenon (see Figure 6.8) is built upon the golden ratio, while in Figure 6.9, an illustration of Paris' Eiffel Tower, the golden rectangle plays a prominent role.

And indeed, the proportions of faces and bodies of many beautiful people have golden rectangles embedded (see the exercises at the end of this section).

Figure 6.8: Greek Parthenon, with golden rectangles (and associated squares) shown in red.

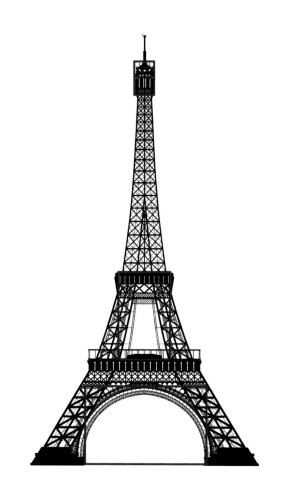

Figure 6.9: Illustration of the Eiffel tower, with golden rectangles shown in red.

There need be no separation between mathematics and art—indeed, what we often love about art is the perfect mathematics underlying the oil paint.

The Eccentric and Symmetric Artist

And the role of mathematics in art doesn't end with proportions. To describe other ways mathematics influences our appreciation of art, we need to talk about **transformations**, which are nothing more than functions, but here they act on whole objects, rather than just numbers. But the idea is the same. The details aren't as important as the ideas, so we'll focus on the latter.

Suppose we start with an object in a picture, say of a bird.

One transformation is called a **translation** (see Figure 6.10), where we simply move the object (here and elsewhere we show the original object slightly transparent in order to focus on the transformation).

Figure 6.10: Translation.

A second type of transformation is called a **rotation** (see Figure 6.11), where we rotate the object about a point (here I rotated the bird around one of the toes).

The third type of transformation we'll consider is called a **reflection** (see Figure 6.12), where we reflect the object about a line (here I reflected the bird around a vertical line running through it).

We can not only apply one of these transformations to an object, but we can do them one after another, for variety's sake.

Figure 6.11: Rotation.

Figure 6.12: Reflection.

For example, if we translate the bird to the right, rotate it, reflect it again, and translate it again, we get the picture in Figure 6.13 (the original and intermediate birds are slightly transparent, and arrows indicate the order of the transformations):

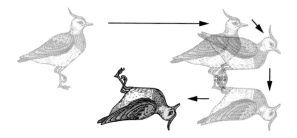

Figure 6.13: Composition of the three transformations.

Doing one transformation after another is mathematically called **composition** and yet again allows for greater variety while maintaining consistency. A **glide reflection** is a translation (that is, a *glide*, followed by a reflection (see Figure 6.14), and figures prominently in many artworks. Figure 6.15 shows a picture created from the picture of the bird, by simply applying the transformations repeatedly to copies of the bird. Does the overall picture seem interesting to you?

Not only can transformations be composed—done successively—but they also can be reversed. The reversal of a translation is a

Figure 6.14: Glide reflection.

Figure 6.15: Creating a picture with translations, rotations and reflections.

translation back to the original position, the reversal of a rotation is a rotation by the same angle in the opposite direction (if the original rotation was clockwise, the reversal is counterclockwise and vice versa), and the reversal of a reflection is the same reflection. These properties make all of the transformations into a mathematical structure called a **group**.

There are, of course, many other types of transformations, such as compressions and stretches, which make scaled versions of the object (see Figure 6.16 for some examples of these). These also are attractive to use and very helpful to artists when creating three -dimensional pictures, as closer objects appear bigger and farther objects always appear smaller, so that when we see compressed and stretched copies of an image, we tend to assign distances to them automatically. When you look at Figure 6.16, don't you see the smaller birds as further away and the big bird on the right as closest?

But compressions and stretches differ in a substantial way from the first three types. Translations, rotations and reflections don't change the figures at all—the objects are exactly the

same size as before. Mathematically, translations, rotations and reflections (and any transformations you get by repeatedly applying them) are called **isometries**, as they preserve the distance between points (that is, if two points have a distance of d between them, their images under an isometry will have the same distance d between them).

And that takes us into symmetry. A **symmetry** of an object is a transformation that takes the object right back on top of itself, so that the image is *identical* to the original. Every symmetry is an isometry, and the symmetries of an object form a group as well (a **subgroup** of all the isometries). For example, in Figure 6.17, if you rotate the square at left about its center by 90°, 180° or 270°, you end up right back with the same square, in the same place (you also do if you rotate the square by 360°, which is the same as not rotating it all—<u>not</u> moving the object is always a symmetry). You can also reflect the square around four lines to have it land on top of itself—the vertical line in the middle, the horizontal line around the middle, and each of the two lines through opposite corners. In Figure 6.17 I labeled the corner points of the original square at left and showed where each of these corners moves to under the corresponding rotation or reflection (lines of reflection, that is the "mirrors," are shown as well). These eight are *all* of the symmetries of the square.

We can compose these symmetries—do them successively, one after the other—to get others. Objects that have a lot of symmetries are often considered attractive, at least to many people,

Don't confuse symmetries as a transformation with symmetry as a relation between objects, as defined back on page 17; they are different. One difficulty with reading mathematics is that sometimes the same term has two different meanings in different contexts, and it is up to us to figure out which one is meant (just as the same word can have more than one meaning, such as "date" or "page.")

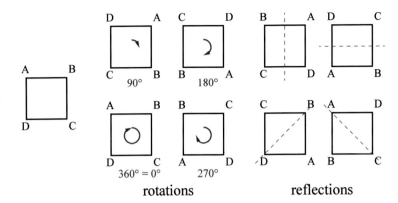

Figure 6.17: Objects with symmetries.

rotations   reflections

although it is quite individual. Remember the inverted U Wundt curve for aesthetics? We tend to like objects that are complex but not too complex. A large number of symmetries tend to make an object easy to remember (and compress—you only need to remember a small part plus the symmetries to generate it). For instance, once you have a single side of the square, the four rotations of the line (about a suitable point) will give you the whole square. On the other hand, too many symmetries can make an object rather mundane and not so interesting. Circles, for example, have infinitely many symmetries—you can rotate them about the center *any* number of degrees and land them right back on top of themselves, and, while perfect, a circle can be boring.

Some objects in real life have symmetries, and this in part makes them attractive objects in artwork. Starfish have reflections and rotations (by multiples of $360°/5 = 72°$) symmetries. Humans (and other animals) essentially have reflection symmetry, about a vertical line down the center of the body—the left and right sides are mirror images of one another. In fact, this symmetry is only approximate—everyone has some differences between their left and right sides. Experiments have shown that those whose faces and bodies are more symmetric tend to start romantic relationships earlier. On the other hand, Lyle Lovett's face is far from symmetric, and he managed to hook beauty Julia Roberts, at least for a while. Imperfection will be attractive, to various degrees, again an illustration of individual Wundt curves

Figure 6.18: Musician Lyle Lovett.

finding the right mix for each of us of meeting expectations and surprise.

> *Life Lesson:* While we find symmetry attractive, we are also attracted to a modest amount of asymmetry as well!

One area that has attracted some artists is **tilings** of the plane. That is, the creation of objects, called **tiles**, such that copies of them (via translations, reflections and rotations) fill up the two-dimensional plane, with no two objects overlapping and no uncovered spots. For example, we can tile the plane with squares—see Figure 6.20—and regular hexagons (that is, six-sided polygons where all sides are the same length)—see Figure 6.21. In Figure 6.20 and Figure 6.21, of course, I only show part of the tiling, but it should be clear that it could theoretically cover the entire infinite plane if extended. And I took artistic license and colored the tiles so that adjacent tiles are different colors—it makes the tiling so much prettier! Both of these have symmetries, but more about that soon.

Figure 6.19: Bees working on a honeycomb. Notice how the side of the honeycomb is a tiling with hexagons—even nature loves tilings!

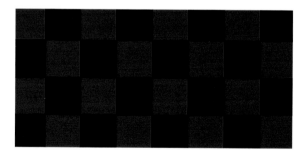

Figure 6.20: Tiling the plane with squares.

There is a whole theory of tilings of the plane, especially those with just a few different tiles. Artists can make great use of tiles. For example, suppose I start with a rectangle?

Figure 6.21: Tiling the plane with regular hexagons.

I next remove the left side, replacing it by a freehand line joining the two left corners.

Next I replace the right edge of the rectangle by a copy of the new left edge.

This ensures that when I translate the tile to the right, it will fit perfectly. I repeat the process with the top and bottom edges, and color in my tile as I like.

Finally, I copy my tile, moving it to the right so that it perfectly fits with the original tile. I am free to redraw or recolor the interior of my second tile as I like (here I just recolored it).

Finally, I recopy and translate my tile all over the plane—they are designed to fit perfectly, vertically and horizontally. I have some choice, and I choose to move the following row over by one tile, so that the coloring is **perfect**—adjacent tiles get different colors (see Figure 6.23). Indeed there are many fish in the sea!

This tiling has two basic symmetries—a horizontal translation and a vertical symmetry (which is a diagonal one, if you want to maintain the colors as well). With some alterations to the process, you can create other kinds of symmetries. For example, starting with a tile (see Figure 6.22) with the tops and bottoms being reflections about their middles, I get the tiling in Figure 6.24, which, in addition to translations, has glide reflections too, with fish in alternate rows swimming against the flow. The one thing an artist always wants is good ideas, and I think these tilings are good food for thought!

You can use simple painting or drawing programs to create fun tilings like those in Figure 6.22—see www.jasonibrown.com/mla for instructions.

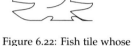

Figure 6.22: Fish tile whose tops and bottoms can be reflected.

Figure 6.23: Tiling the plane with fish!

Figure 6.24: Another tiling of the plane with fish and glide reflection symmetries.

Exercises

Exercise 6.1.1. Which of the following rectangles are golden rectangles? Explain your answer.

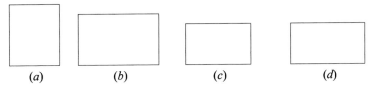

(a)   (b)   (c)   (d)

Exercise 6.1.2. The most common size of index cards in North America is 3 inches by 5 inches. Why do you think these dimensions are aesthetically pleasing?

Exercise 6.1.3. Find the length of line segment BF in Figure 6.4.

Exercise 6.1.4. Why does the circle in Figure 6.4 pass through point A?

Exercise 6.1.5. Find the length of line segment DX in Figure 6.4.

Exercise 6.1.6. From the formula $\phi^2 = \phi + 1$, show that $\phi = \frac{1+\sqrt{5}}{2}$ since ϕ is bigger than 1.

Exercise 6.1.7. From the formula $\phi^2 = \phi + 1$, show that $\frac{1}{\phi} = \phi - 1$, which is about 0.618.

Exercise 6.1.8. Find decimal approximations to the numbers

$$1, 1+1, 1+\frac{1}{1+1}, 1+\frac{1}{1+\frac{1}{1+1}}, 1+\frac{1}{1+\frac{1}{1+\frac{1}{1+1}}}.$$

What do they seem to approach?

Exercise 6.1.9. Calculate the 10th Fibonacci number and compare it to $\frac{\phi^{10}}{\sqrt{5}}$. What do you find?

Exercise 6.1.10. For the following picture of actor Mark Wahlberg, find golden rectangles that fit his face, torso and legs.

Exercise 6.1.11. Translate the following figure to the right 2.5 inches.

Exercise 6.1.12. Translate the following figure to the right 2.5 inches and down 1 inch.

Exercise 6.1.13. Rotate the following figure to the right 90° about its center.

Exercise 6.1.14. Reflect the following figure along a vertical line right down the center of the figure.

Exercise 6.1.15. Reflect the following figure along a horizontal line right down the center of the figure.

Exercise 6.1.16. Describe a sequence of translations, rotations and reflections that will carry the original drawing on the left into the one on the right.

Exercise 6.1.17. Describe a sequence of translations, rotations and reflections that will carry the drawing on the right into the one on the left.

Exercise 6.1.18. Is there an isometry that carries the figure at the left into the one on the right? Explain.

Exercise 6.1.19. Is there an isometry that carries the figure at the left into the one on the right? Explain.

Exercise 6.1.20. What symmetries does an equilateral triangle have? What about an isosceles triangle that is not an equilateral triangle?

Exercise 6.1.21. What symmetries does a starfish have?

Exercise 6.1.22. Show that you can tile the plane with equilateral triangles.

Exercise 6.1.23. Describe the symmetries of the tiling in Figure 6.23. Does the answer change if you ignore the colors of the tiles?

Exercise 6.1.24. Describe the symmetries of the tiling in Figure 6.24. Does the answer change if you ignore the colors of the tiles?

Exercise 6.1.25. (Harder) The sum of all the interior angles (in degrees) of a polygon with n sides ($n \geq 3$) is $180(n-2)$.

(a) Show that each angle of a regular polygon with n sides is $\dfrac{180(n-2)}{n}$.

(b) If you can tile the plane with a regular polygon with n sides, then some number, say k, of them meet at a point. Show that this means that

$$360 = k\frac{180(n-2)}{n}.$$

(c) Conclude that $k = 2 + \frac{4}{n-2}$, and that the only possible values of n for a tiling with a regular polygon with n sides are $n = 3, 4$ and 6 (that is, the only possible tilings with a regular polygon are the ones found for triangles, squares and hexagons).

6.2 Believing What You See (or Not)

We've seen how we can use transformations (compression and stretches) to create the illusion of distances. But mathematics provides even more in the way of guidance in producing the illusion of three dimensions in paintings and drawings. Prior to the Renaissance, the only visual clues used to show distance in artwork was to change the size of objects (smaller ones being further back) or by overlapping images. But starting in the 1400s, artists began to create more realistic three-dimensional paintings, using geometry, with much of the work being done in Italy.

It's All a Matter of Perspective

Proper perspective depends on the notion of infinity in an essential way. How does one create the illusion of the third dimension in a two-dimensional artwork? The simplest process involves setting a point in the painting as **point at infinity**, also called the **vanishing point**, set off along the horizon. The key is that if you observe parallel lines that head off into the distance, they seem to converge to a point, even though we have been taught since we were small that parallel lines *never* meet. Giving up this throwback to Euclidean geometry allowed artists in the Renaissance to finally figure out how to render the third dimension in their art.

Here is an example of artwork whose perspective is generated by a vanishing point (shown in red on top of the painting).

vanishing point   horizon line

Once the point at infinity is set, one makes all lines in the picture that are perpendicular to the plane of the canvas, that is, all lines that head into the picture, going from the foreground all the way back into the background, meet at the point at infinity; lines that are parallel to the plane are shown in the usual way. So, for example, if I want to draw a picture of train tracks heading into the distance, I start with the rails. As the rails are perpendicular to the plane of the picture, that is, they head back into the distance, they, though they are parallel in real life, should both converge to the point at infinity.

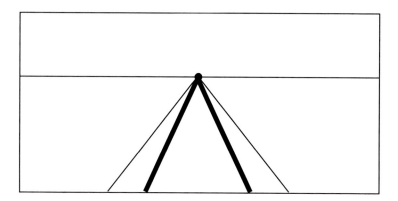

Now I add the railway ties across the rails. The sides (but not the fronts) of the ties are perpendicular to the plane of the picture, so the lines formed from the edges will also head toward the vanishing point, while the fronts are horizontal, and subsequently the ties get smaller as they move off into the distance.

I now add telephone poles, which appear vertical as usual, though the tops will form a line perpendicular to the plane of the picture and hence back toward the vanishing point.

Finally, I add in some telephone wires, a sun at the horizon and some sun rays. Of course, this is only the start, but I think you can see how the vanishing point adds greatly to the perspective. I think we are on track!

I should mention that artists have used variations with multiple (two or three) vanishing points. Figure 6.25 shows a painting by painter Johannes Vermeer with two vanishing points (one to the left, one to the right, off the canvas, as indicated by the red lines extending the checkerboard floor's lines).

Fractals and Pictures within Pictures

The drawing I created used simple shapes—lines, rectangles and circles—but such simple shapes are not what most people

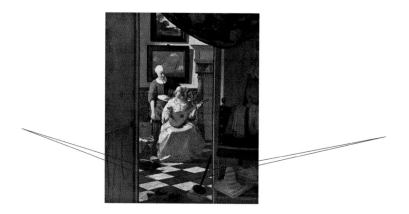

Figure 6.25: Painting by Dutch artist Vermeer with two vanishing points.

consider "beautiful." What do people like to look at? Not simple geometric objects, but ones from real life, such as trees, mountains, coastlines, and so on. The latter are not easy to draw with the standard geometric objects like lines, circles and polygons, but they can be created using different mathematics.

One way to create a coastline is to start with a simple figure, like the one below, made up of just four equal length line segments; the middle two make an angle of 120° with the horizontal lines (they would form an equilateral triangle; if you added in the missing middle line segment).

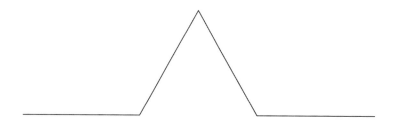

Figure 6.26: A fractals' humble beginning—the original simple curve.

Then you proceed to build your coastline as follows. Replace each of the four line segments by a smaller copy of the original figure, as in Figure 6.27.
We continue with this figure, replacing each of the sixteen line

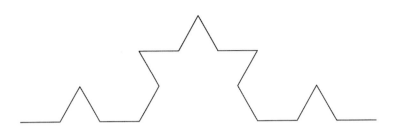

Figure 6.27: Curve's first replacement.

segments with smaller versions of the original. Figure 6.28 shows what you get after a few more repetitions of the process. Doesn't this remind you of a coastline?

Figure 6.28: Fractal coastline.

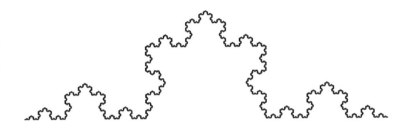

The figure you get in the limit, that is, theoretically after doing infinitely many such replacements (in practice the figure will not change much after several iterations, that is, repetitions of the process) is an example of a **fractal** (the specific one illustrated in Figure 6.28 is called the **Koch curve**). The idea is that just from a small beginning, the process will generate objects that often have a natural look to them.

Fractals have a property called **self-similarity**—the figure is made up of several scaled versions of the whole picture. The picture in Figure 6.28 is made up of four scaled versions of the figure itself. It seems so unintuitive and paradoxical—having an object made up of smaller versions of itself—but these kinds of fractals are not only all around us, but they are among the objects we find most beautiful, perhaps exactly for their self-similar properties!

324

And the process is even more revealing. Suppose each of the line segments in the original simple curve in Figure 6.26 has length 1, so that the curve has length 4. Then in each of the four scaled versions for the next replacement, each line segment has length 1/3, so we replace each of the four original line segments with a figure that has length $1/3 + 1/3 + 1/3 + 1/3 = 4/3$. The length is now $4(4/3) = 16/3 \approx 5.3$, or 4/3 times the original length. The next repetition will have length 4/3 times that of the previous one, so length $4(4/3)(4/3) = 4(4/3)^2 \approx 7.1$. At each iteration, you replace each line segment by a figure with 4/3 times the length. The lengths of the figures are therefore

$$4, 4(4/3), 4(4/3)^2, 4(4/3)^3, \ldots$$

As these numbers get as big as we like, the length of the curve gets huge, even though it's in a confined space. This is why it is impossible to accurately measure the length of a real-life coastline accurately—it is very dependent on how you measure it.

You can build a two-dimensional fractal in a similar way. For example, suppose you start with an equilateral triangle and punch out the middle triangle:

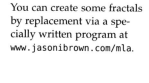

You can create some fractals by replacement via a specially written program at www.jasonibrown.com/mla.

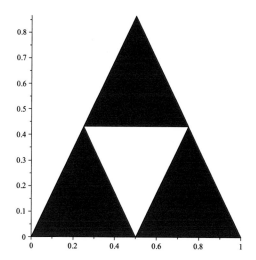

Repeating the process, punch out the middle triangle in each of the remaining three equilateral triangles, and so on.

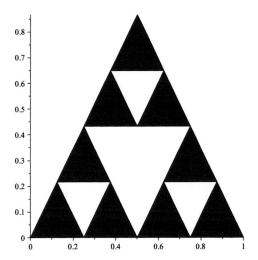

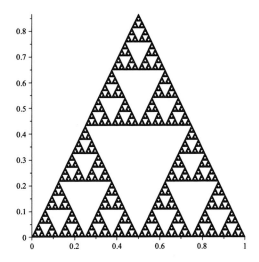

Figure 6.29: Sierpinski's triangle after six iterations.

Each time you remove 1/4 of the area that was there previously, leaving 3/4 of what was there, so if the area of the original triangle was A, then the areas give you the sequence

$$A, \frac{3}{4}A, \left(\frac{3}{4}\right)^2 A, \left(\frac{3}{4}\right)^3 A, \dots$$

which clearly get closer to 0. So the fractal that results, called **Sierpinski's triangle**, has no area at all! It does form a pretty picture, though, as does a similar kind of fractal shown in Figure 6.30 where you cut up squares into nine equal subsquares

and repeatedly punch out the middle one—wouldn't it make a
nice pattern for a carpet?

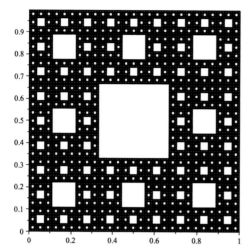

Figure 6.30: Sierpinski's
carpet after six iterations.

Fractals abound in life. Realistic pictures of lightning bolts,
clouds, mountains and trees can all be generated via fractals.
Figure 6.31 starts with a small figure on the left, with just seven
straight line segments, a "stick" figure tree a child might draw.
The replacement is carried out repeatedly on every line segment
except the bottom one (the "trunk" of the tree). After several
repetitions, you end up with the figure on the right. Not a
perfect tree, but certainly a good approximation!

Many other objects in nature (even in three dimension, instead
of just one or two) are fractals as well. For example, a stalk of
broccoli is a fractal—if you peeled off a part and examined it,
you would see it looked like a small whole stalk of broccoli. And
even within our body there are fractals, with fractal properties
that are vital to our lives. For example, the branching structure
in our lungs is fractal—we have large main bronchial tubes, with
branches off them, and branches off the branches, and so on.
Just as the Koch curve had unbounded length even though it
didn't get too large, the fractal structure of our lungs allows for
a huge amount of surface area to exchange oxygen, while being
in a relatively small space like our chests. Without this fractal
property (large surface area in a confined space) we wouldn't be
able to get enough oxygen to live.

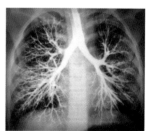

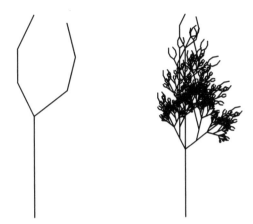

Figure 6.31: Fractal tree.

Seeing into Infinity

Fractals, with their self-similarity, can often seem paradoxical. We have seen examples of fractals that are in a confined area and yet have unbounded length, and ones that seem to take up space but lack any area! Figure 6.32 shows a doorway that contains a doorway which contains a doorway, and so on. Its fractal nature draws us in and suggests an infinite number of entrances, leading on forever. Is this disturbing or just interesting?

Fractals have infinity built right into them—the fractals themselves are the objects that arrive after *infinitely* many steps, though the pictures we generate of course are only approximations, stopped after some finite number of iterations. Our eyes are limited as to how small a detail we can see, but the pictures of fractals allude to infinite progressions of smaller objects, and that glimpse into the infinite is appealing.

Another way of portraying infinity is with infinitely climbing objects. These visual paradoxes cycle around, climbing higher, and yet return to where they start, in an endless cycle. Follow the water around in Figure 6.33 and watch how it continually rises!

We'll show one more visual paradox, a picture of a three-dimensional one, called the **Necker cube**. Have a look at Figure 6.34— do you see how it really can't exist in space, though it seems to?

Figure 6.32: Fractal entrance-way to the infinite.

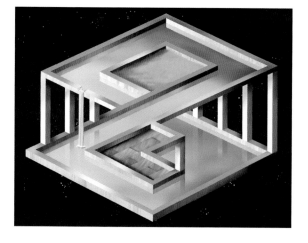

Figure 6.33: Infinitely climbing waterfall.

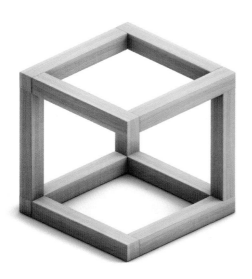

Figure 6.34: The impossible Necker cube.

In the last chapter we'll return to visual paradoxes. But I'm going to take the last part of this section to show you how pictures can even be *proofs* of mathematical statements, which may seem paradoxical in itself, as aren't proofs by necessity done with words and formulas? Not at all—proofs are convincing arguments, nothing more, and nothing less. How the convincing is to be done can vary—it can be written with text or formulas, heard in a lecture, or seen with a picture. And the last of these is what I want to illustrate here.

For example, what is

$$1 + 2 + 3 + 4 + \cdots + n$$

that is, the sum of the first n whole numbers? For example, the sum of the first four whole numbers is

$$1 + 2 + 3 + 4 = 10.$$

What we are after is a formula. To get the formula say for $1 + 2 + 3 + 4$, look at a 4 by 5 array of dots:

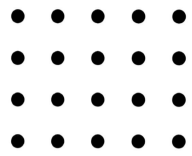

All I'm going to do now is color the first dot in the first row, the first two dots in the second row, the first three dots in the third row, and the first four dots in the fourth row:

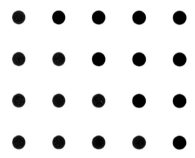

The number of red dots is $1 + 2 + 3 + 4$, exactly what we want to add up. But this is the same as the number of black dots, as the collection of black dots is just the collection of red dots, rotated around the middle. Just look at the picture! So there are exactly the same number of black dots as well, $1 + 2 + 3 + 4$. But the total number of dots of either color is obviously $4 \times 5 = 20$, as there are 4 rows of 5 dots. The red dots comprise exactly half of these, and hence there are $20/2 = 10$ red dots, that is,

$$1 + 2 + 3 + 4 = 10.$$

More generally, we can draw a *general* picture for adding up $1 + 2 + 3 + \cdots + n$, with an array of n by $n + 1$ dots:

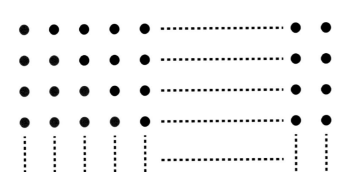

I started by coloring in the dots as in the previous argument, 1 red dot in the first row, two in the second, all the way down to n in the last row (leaving one black dot). But then again we see that the number of red and black dots is the same (it is visually obvious!), so that the number of red dots is half the total, which is clearly $n \times (n + 1)$. So we have a formula and a proof:

$$1 + 2 + 3 + \cdots + n = \frac{n(n+1)}{2}.$$

For $n = 4$, we get $1 + 2 + 3 + 4 = 4(4 + 1)/2 = 10$, as before.

This argument is (at least to me, and I hope to you) completely convincing—a picture proof! Of course, you could write out an algebraic proof, but why would you—the proof in the picture is so clear, and dare I say, so beautiful! And the result is an infinite number of statements that we have proved, one for each whole number n.

I'll end with one picture proof of a formula for the infinite geometric series

$$\frac{1}{2} + \frac{1}{4} + \frac{1}{8} + \cdots.$$

We saw how to add it up algebraically in Section 4.1 when we were considering perpetuities, but we don't need algebra to find the formula or even prove it. Consider the following red cake:

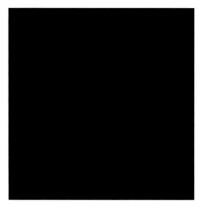

We continue by eating 1/2 the cake (the piece removed is shown in black):

We continue by eating half of what is left, which is half of 1/2, or 1/4 of the cake:

Again we eat half of what is left, which is 1/8 of the cake:

If we continue on forever, how much of the cake do we eat? On one hand, we eat

$$\frac{1}{2} + \frac{1}{4} + \frac{1}{8} + \cdots$$

in total. But clearly we eventually eat every part of the cake, so we eat one whole cake! Thus

$$\frac{1}{2} + \frac{1}{4} + \frac{1}{8} + \cdots + 1,$$

an infinite sum we have seen before, but now with a much tastier proof!

Exercises

Exercise 6.2.1. How many vanishing points does the following painting have? Mark them.

Exercise 6.2.2. How many vanishing points does the following painting have? Mark them.

Exercise 6.2.3. Draw a mountain range in perspective with a single vanishing point.

Exercise 6.2.4. Draw a highway with cars in perspective with a single vanishing point.

Exercise 6.2.5. Suppose each of the line segments in Figure 6.26 has length 5. What is the length of the curve in Figure 6.27?

Exercise 6.2.6. Explain why the area under the Koch curve (but above its horizontal base) is bounded.

Exercise 6.2.7. Do you think there are any smooth parts to the Koch curve? Explain your answer.

Exercise 6.2.8. Use a round pie instead of a square cake to give a picture proof that

$$\frac{1}{2} + \frac{1}{4} + \frac{1}{8} + \cdots = 1.$$

Exercise 6.2.9. Consider a square array of n by n dots, as shown below. Add up the reverse L-shaped areas of dots. What does this picture prove?

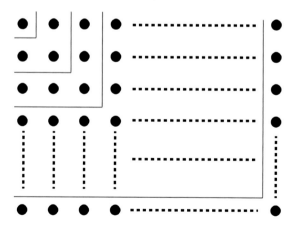

And Now the Rest of the Story . . .

The door that MC Escher walked through was the mathematical world of geometry. He couldn't understand the language in the paper—the mathematics was way too theoretical for someone unschooled in proofs— but Escher had an innate understanding of the geometry from the figures included.

Escher devoured the paper's figures, and he was addicted to their study. How could he make use of them? He had already in parallel been carrying out his own studies, his own research, in tilings, finding out rules and principles for generating them, but now he had some fellow devotees, though they were mostly interested in the math, not the art. Escher continued with his "addiction" to tilings, but his work was becoming more and more abstract all the time. Early in his career he was moved to capture landscapes, but more and more the pictures he attempted to capture were not in the world but in his mind.

1954 was a tipping point. The International Congress of Mathematics (ICM) was held in Amsterdam, and Escher, already renowned in art circles for his expressive and captivating artwork, gave a speech, even though he was an artist, not a mathematician. Though Professor H.M.S. (Harold Scott MacDonald, or just "Donald") Coxeter, a world famous geometer from the University of Toronto in Canada had missed the talk, he bought two of Escher's prints. After writing to Escher to ask to reproduce two of his artworks for a geometry paper he was writing, Coxeter mailed back to Escher a copy of his latest geometry paper, on usual geometries on spheres and something called the hyperbolic plane.

The inspiration was instantaneous and ground-shaking. In Coxeter's paper (among the math jargon, the "hocus-pocus text" [1] that Escher couldn't make sense of) there were drawings of infinitely smaller geometric objects on the surface of a circle, ones that seemed to approach infinity as they decreased in size. Escher expended a huge effort to understand the pictures, how they were created, until he could make the process his own.

In subsequent years, Escher and Coxeter exchanged a lot of correspondence, both geometers, though one an artist, the other a mathematician. Coxeter appreciated Escher's natural talents and intuitive understanding of geometry and was always willing to share his knowledge, the consummate educator that he was. Escher stated that the exchanges changed dramatically the direction of his art, art that still enthralls viewers and artists—artwork that might not have existed without the mathematics.

Figure 6.35: M.C. Escher (1898–1972). All M.C. Escher works ©2014 The M.C. Escher Company, The Netherlands. All rights reserved. Used with permission. www.mcescher.com

[1] S. Roberts. *King of Infinite Space*. Anansi, Toronto, 2006

6.3 *Review Exercises*

Exercise 6.3.1. Which of the following rectangles are golden rectangles? Explain your answer.

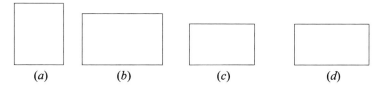

(*a*)   (*b*)   (*c*)   (*d*)

Exercise 6.3.2. For the following picture of actress Julia Roberts, find golden rectangles that fit her face, eyes and nose. What do you find about the dimensions of her mouth?

Exercise 6.3.3. Add up the first 10 Fibonacci numbers and compare it to the 12th Fibonacci number, f_{12}. What do you find?

Exercise 6.3.4. Rotate the following figure 180° about its center.

Exercise 6.3.5. Reflect the following figure along a vertical line right down the center of the figure.

Exercise 6.3.6. Describe a sequence of translations, rotations and reflections that will carry the original drawing on the left into the one on the right.

Exercise 6.3.7. What are the symmetries of the following polygon?

Exercise 6.3.8. How many vanishing points does the following painting have? Mark them.

Exercise 6.3.9. If you start with a square with area 1 and punch out the middle square as in Figure 6.30, what is the area after one step? After two steps? Three?

Exercise 6.3.10. Explain how the following gives a picture proof of

$$f_n^2 + f_{n-1}^2 + \cdots + f_1^2 = f_{n+1}f_n.$$

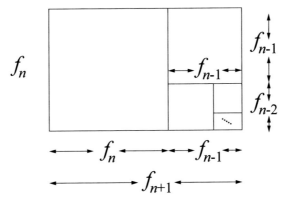

Mathematics of Sound (and the Sound of Mathematics)

It was the early afternoon on April 16, 1964, and The Beatles were working on what would be the opening song on the next album and their first movie. John had eagerly taken on the task of composing the number, beating Paul in the friendly competition of who would compose it. The title had been a phrase out of the mouth of their drummer Ringo, who had blurted out one evening after a concert, "it's been a hard day's night." Ringo was always saying stuff like that, with no intention of being funny, but what a great line it was! The cheekiness of the phrase immediately caught John's attention, and in short order he had the framework for the song worked out (with lyrics written on the back of a birthday card sent to his son Julian by a fan).

The Beatles honed the song through a number of takes in Abbey Road studio, and they were getting ready for one final one, take 9. Their producer George Martin was keen on getting a "big" opening chord to start the song, as it would open not only the album but their feature film. The chord evolved throughout the morning and early afternoon, but wasn't quite there (like George's solo, which would have to be set aside for another track later). Martin pushed the button on the recording console and asked for one final run-through. As John started to count in, the producer realized that he had wanted them to retune their instruments, but lunch was waiting and enthusiasm was waning, so he just decided to let it go. On with the show!

Take 9 was fantastic, all thought, though a little work still needed to be done later—Harrison needed to fill in his solo, which he heard in his head but hadn't figured out how to play it perfectly yet, and the opening chord still needed something. Martin planned to come in later with just the sound engineer to add a bit, just to make the opening chord grander, and perhaps a bit of a mystery. But everyone knew that things would never be the same. Yeah, yeah, yeah!

EVERYONE LOVES MUSIC. The best songs have the power to move you emotionally, from the depths of despair to the heights of ecstasy. You might think that nothing could be further from music than mathematics, but you would be wrong, wrong, wrong! Musicians use more mathematics when they play, record and write than practically any other professional. And the music we find most intriguing to listen to music that is filled with interesting and wonderful patterns. Mathematics has often been called the science of patterns, and with a wee bit of mathematical knowledge, you can better appreciate the music you enjoy.

7.1 *Mathematics of Listening*

It is quite amazing how our perception of music is so intrinsically mathematical. And that does not depend on age, gender, culture or even mathematical ability. We humans are just built to appreciate the math in music, and that appreciation infuses both our listening and composing of music. Naturally, listening and composing influence one another, but to begin, let's examine how we listen to music. You'll see how integrated mathematics is in the development of basic musical elements like notes, scales and rhythm. And you'll learn some details about what notes and scales are, and how they developed. Even if you are a seasoned musician, the mathematical view of music will teach you some new tricks.

"Trigging" on to how we hear

We can't separate out mathematics from hearing. Every sound is caused by the vibration of air molecules moving back and forth in quick succession, and our ears and minds are built to recognize and respond to such vibrations. Musical sounds are made up of *pure tones*, which correspond to objects vibrating in a wave-like pattern.

The wave for a pure tone is exactly the **sine** (and **cosine**) curve you saw back in high school. If only teachers had told you back then that trigonometry was the basis for all music! Listening to music depends deeply on our inherent ability to

You will find relevant musical examples and useful musical software for this chapter at www.jasonibrown.com/mla.

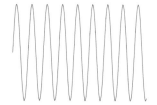

Figure 7.1: Simple sound wave.

recognize and appreciate trigonometric functions, whether we are mathematical or not.

Two musical properties of pure tones are directly related to the mathematical properties of their sine waves:

- The *loudness* of a pure tone corresponds approximately to the sine curve's **amplitude**—how high and low it goes vertically from the midpoint. The higher the curve goes, the louder it seems. Figure 7.2 shows a sine curve with amplitude 2.

- The *pitch* of the tone corresponds to the **frequency** of the curve, that is, how often it repeats. This is measured in *hertz* (Hz), the number of cycles per second that the sine wave repeats (a *kilohertz* (kHz) is a thousand cycles per second). Figure 7.2 shows a sine curve with frequency 5 Hz (as it repeats five times every second.

The range for human hearing is 5 Hz to 20,000 Hz (that is, 20 kHz), though as we age, we tend to lose hearing at the top end. Other animals have different ranges. For example, dogs can hear in the range of 40 Hz to 60 kHz, while some bats can hear frequencies around 200 kHz.

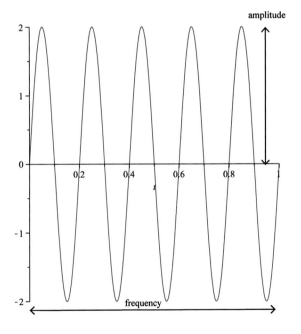

Figure 7.2: Sine curve, with amplitude and frequency indicated.

Mathematically, the pure tones correspond to functions like

$$A \sin(2\pi \times f \times t),$$

where A is the amplitude, π is our good friend from the area of a circle, f is the frequency in hertz, and t is time (in seconds). So, for example, the function $1.3\sin(2\pi \times 330 \times t)$ has amplitude 1.3 and frequency 330 Hz, while the function $0.7\sin(200\pi t)$ has amplitude 0.7 and frequency 100, as we can rewrite it as $0.7\sin(2\pi \times 100 \times t)$.

People don't tend to like to listen to pure tones—they sound too simple. All musical tones are made by adding together such sine curves, and these combinations give us the sounds we hear, from the sweet texture of a flute to the rough crunch of a rock guitar chord. By combining sine curves we get repeating patterns that not only our ears can appreciate, but our eyes as well. Figure 7.3 shows an example of the sum of a few such simple sine curves. The resulting function repeats, but its shape is fascinating.

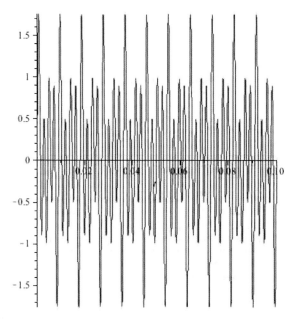

Figure 7.3: Plot of $0.5\sin(2\pi \times 220 \times t) + 0.4\sin(2\pi \times 330 \times t) + \sin(2\pi \times 440 \times t)$. The mathematical function does indeed have a sound!

Piano tuners inherently use trigonometry to tune pianos! In a piano, there are three strings under each hammer (but not all, more about this later) that ring at the same frequency when a key is struck. I say at the same frequency, but it is up to the tuner to tune them identically. Once the piano tuner has tuned one of the strings accurately, he then plucks this string with one of the other two and listens carefully. If the string is slightly out of tune, he hears a single frequency (rather than two) that warbles in a "wah-wah" effect that is called beats. He then tunes the second string, listening to the beats, which slow down as the two frequencies get closer together, and the beats disappear entirely when the two strings are at the same frequency.

Why does this effect happen? Remember we said that we humans can hear combinations of sound waves, which are sine curves. Back in high school, you were no doubt tormented with proving **trigonometric identities**, which were little results that said that one expression involving trig functions was always equal to another, seemingly different looking one. For example, you might have been asked to prove that $2\sin^2 x - 1 = 1 - 2\cos^2 x$. Here is another trig identity:

$$\sin A + \sin B = 2\cos\frac{A-B}{2}\sin\frac{A+B}{2}.$$

This is probably one you haven't seen before, but take it from me, it is always true, for any numbers A and B. Now in the case when A and B are close together, as when two frequencies are close but not equal, instead of hearing two notes (which is what the left side of the trig identity expresses) we hear a single note at the average frequency (which is the last term on the right side). The "wah-wah" effect is produced by the cosine term, which disappears when A equals B, that is, the two frequencies are identical. We cannot but help to hear this trigonometric identity. Beats give instruments like pianos and 12-string guitars their warmth, and they are one of the reasons we like to listen to choirs—it's not so much the perfectness of tunings, but the small amount of "out-of-tune-ness" that we like so much! In fact, many recording artists record their voices twice (known as "double-tracking" their vocals) for this reason.

Fractions, powers and scales

The first to consider what pairs of musical tones were aesthetically pleasing (that is, *consonant*) were the ancient Greeks, who placed music on the same pedestal as mathematics. Fretting a string (that is, placing a finger on the string so as to cut down

In the computer program "Trig Sounds," found at www.jasonibrown.com/mla, assign a frequency of 120 Hz to one note and 126 Hz to the second. Listen for the beats. Then change the frequency of the second note, bringing it closer to 120 Hz, and listen to what happens to the beats.

A *trigonometric identity* is an equation involving trigonometric functions (like sines, cosines and tangents) that is true in all instances.

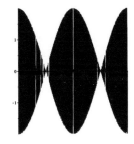

Figure 7.4: Not only can you feel the beat, now you can *see* it!

the length of the string vibrating when plucked) exactly halfway between the top and bottom produces, when plucked, a tone that is exactly an *octave* higher, that is, at twice the frequency of the original note. The connection between a note and an octave above is so strong, across all cultures, that men and women often sing an octave or two apart when trying to sing the same note. We call a set of frequencies that are some number of octaves away from one another a *pitch class*, and the ratio of two such frequencies is always a power of 2.

Two notes, an octave apart, always sound good together, though perhaps a little boring. Pythagoras (the mathematician who is famous for his $a^2 + b^2 = c^2$ theorem on the sides of right-angled triangles) discovered that ratios of small fractions were the key to nice sounding intervals (an *interval* is a pair of notes struck together or quickly, one after the other). Two frequencies, one that is 3/2 times the other, is what we call a *perfect fifth*. It is the basis of rock power chords on a guitar. *Perfect fourths* are based on the ratio of 4/3 and occur less frequently in rock music. Our (mathematical!) minds and ears find frequencies that are in ratios of small fractions to be pleasing. We don't pay as much attention to the **difference** between frequencies as to the **ratio**, and the same ratio of two frequencies sounds the same no matter what the frequencies are.

Now as humans we find infinite choice often overwhelming. And there are infinitely many frequencies available to listen to. So what we like to do is to intentionally whittle our choices down to a set of "steps"; this process of *discretization* reduces our choices of frequencies to musical *notes*. One fixed frequency is set—for historical reasons, it is usually a frequency of 440 Hz, and the note is called "A"—more precisely, A4 . The number 4 denotes the octave it sits in—the note up one octave, with frequency 2 × 440 = 880 Hz is called A5, while the note down the octave (corresponding to *dividing* the frequency by 2) is A3, with a frequency of 440/2 = 220 Hz. The often spoken about "middle C" has a frequency of about 262 Hz (more about this shortly). As mentioned earlier, a *pitch class* is a note together with all of the other notes that are some number of octaves up or down; for example, the pitch class A is the collection of

If you have a guitar available, try plucking a string, then touching the string lightly with a finger to cut the string in half (at approximately the twelfth fret) and repluck to hear the octave at twice the frequency. Do the same at the seventh fret (about 1/3 the way down the neck) and at the fifth fret (about 1/4 the way down the neck).

notes some number of octaves up or down from A4, that is, $\{A0, A1, A2, A3, A4, A5, \ldots\}$. Across all cultures, people tend to find that notes in the same pitch class sound "the same" in some sense.

The set brackets { and } enclose a collection of items.

Once one fixed note is fixed, other notes were generated by the Greeks by multiplying the frequencies by small fractions (so that the ratios are small fractions). So, for example, a perfect fifth up from A4, called E5, has frequency $(3/2) \times 440 = 660$ Hz, and the ratio of the frequency of E5 to that of A4 is $660/440 = 3/2$.

E is 5 letters up from A: A,B,C,D,E, hence the reason a perfect fifth up from A is an E note.

Small fractions between 1 and 2 were used to fill in notes between A and the A one octave up, but not all small fractions were used for two reasons—we need to have adequate separation between the notes, and some ratios just don't sound as good as others!

Collections of notes within an octave are called a *scale*. The *Pythagorean scale* (sometimes called the *major scale*), based on eight small fraction ratios (that is, ratios of whole numbers that are small), uses the numbers

$$1, \frac{9}{8}, \frac{81}{64}, \frac{4}{3}, \frac{3}{2}, \frac{27}{16}, \frac{243}{128}, 2$$

so that for a scale starting on middle C4 (262 Hz), we have the frequencies given in Table 7.1. (For example, to get the second note, D, we multiply the starting frequency, 262 Hz by 9/8, and so on.) Not all the fractions might seem small to you, but in the scheme of things, they aren't too large.

C	D	E	F	G	A	B	C
262 Hz	295 Hz	332 Hz	349 Hz	393 Hz	442 Hz	497 Hz	524 Hz

Table 7.1: Frequencies for the notes in the Pythagorean scale, starting at middle C.

As a side "note," music, like mathematics, has developed its own written notation. Notes are often written on a *musical staff,* a series of horizontal lines and spaces that separate them. The treble staff is shown in Figure 7.5, with the notes from middle C up to the following octave shown as ovals (with sticks on one side to indicate timing considerations—more about this later); as the frequency rises, so does the placement on the staff, with consecutive notes occupying an adjacent line and space (or space and line).

You might wonder if it is important to learn to read and write music, as many famous musicians can't write or read musical notation. But both music and mathematics are abstract forms for which a system of notation is vital for communication.

C D E F G A B C

Figure 7.5: The major scale starting at middle C, C4. We'll talk more about musical notation throughout, although a complete discussion is beyond our scope in this text.

Not only are the small fraction ratios of tones nice sounding, but they are built into the physics of how musical instruments produce notes. Typically, every note played on a melodic instrument is made up not only of a primary frequency (also called the *fundamental* tone), but also of other frequencies that are higher than the original tone. These additional tones have frequencies that are whole number multiples of the original frequency. For example, if you pluck an open A string on a guitar, which has a frequency of 110 Hz, then the frequencies 220 Hz, 440 Hz, 660 Hz, and so on are produced. Each melodic instrument produces these different overtones (or *harmonics*); the relative loudness of each gives the instrument its characteristic sound or *timbre* and is one of the reasons we can distinguish a note played on a guitar from the corresponding one played on a clarinet. It is yet again one of the amazing connections between mathematics and music that these harmonics are exact whole number multiples of the original frequency.

On the software program, "Trig Sounds," set one tone at frequency 120 Hz, and set a second tone at different small fraction multiples of this tone (say at 240 HZ, which is twice the frequency, then at 180 Hz, which is 3/2 times the frequency, then at 160 Hz, which is 4/3 times the original frequency). Listen to the sounds. Play around with the amplitudes and then alter the frequencies.

The most pleasing sounding pairs of notes correspond to small fraction ratios. But working with small fractions poses a musical dilemma. We naturally identify notes that are octaves apart as being essentially the same. So we would expect that if you repeated the nice sounding interval of a perfect fifth, eventually we should end up at some octave above the note at which we started, that is, in the same pitch class as the starting note. So starting with a frequency of 110 Hz (a low A), a perfect fifth up would be at frequency $110 \times \frac{3}{2} = 165$ Hz (an E), another perfect fifth up would be at frequency $110 \times \frac{3}{2} \times \frac{3}{2} = 165 \times \frac{3}{2} = 371.25$ Hz (a B note), and so on. We can get pretty close; 12 repetitions of a perfect fifth takes us to a frequency of $110 \times (3/2)^{12} = 14,272$ Hz (remember, every time you go up a fifth, you multiply the frequency by 3/2). This sounds quite close to the seventh octave, A9, above the original A2, which has

frequency
$$110 \times 2^7 = 14{,}080 \text{ Hz.}$$

Close, but no cigar! Could we ever get back to the same pitch class, A, as the original A2 note 110 Hz? Sadly, math tells us no. For if some piling up of perfect fifths, say m of them, ended up in the same pitch class as the original note, say n octaves above the original note, then we would have to have

$$110 \times \left(\frac{3}{2}\right)^m = 110 \times 2^n.$$

Canceling out 110 on both sides we get

$$\left(\frac{3}{2}\right)^m = 2^n$$
$$\frac{3^m}{2^m} = 2^n$$
$$3^m = 2^n 2^m$$
$$3^m = 2^{n+m},$$

Remember that when you multiply powers with the same base, you *add* the exponents.

that is, some power of 3 is equal to a power of 2. But the left side is odd, while the right side is even, so this can never happen - we can never end up in the same pitch class as the original tone, no matter how many perfect fifths we move up.

There are even bigger problems with using small fraction ratios for notes. It turns out that the frequencies of notes in different keys vary depending on how you calculate them, and this makes it practically impossible to change keys in mid-song (changing keys in the middle of a song is called *modulation*, technique that the best composers essential in composing interesting songs).

What was the way out of a musical dilemma? A mathematical solution, of course!

Here is what was proposed (and accepted) back in the 16th century. It was decided to divide an octave into 12 equal steps called *semitones*, with the *ratio* of each step being the same.

Calling the ratio of a semitone r, we see that if the frequency of
the starting note is f Hz, then the frequency of the first semitone
up would be $r \times f$, the frequency of the second semitone (a *whole
tone*) up would be $r \times r \times f = r^2 \times f$, and so on. Going up 12
semitones from some frequency f took you, on one hand, to a
frequency of $r^{12} \times f$ while, on the other hand, ending up at the
octave takes you to twice the frequency, $2 \times f$. So we need to have
$r^{12} \times f = 2 \times f$, which, by canceling out f, implies that $r^{12} = 2$.
This means that the ratio for a semitone must be $\sqrt[12]{2}$, the twelfth
root of 2 (also written mathematically as $2^{1/12}$), which is about
1.06 and is an *irrational* number—although we have come to it
through completely rational means!

The scale of 12 semitones, each corresponding to multiplication of the frequency by $\sqrt[12]{2}$ (which can also be written as
$2^{1/12}$) leads to the *equally tempered scale*. You might think that it is
strange to let mathematics dictate music considerations, and to
be sure, there are some disadvantages to using the twelfth root
of 2 instead of the small fractions. We have lost all the perfect
sounding intervals based on small fractions. A perfect fifth is no
longer multiplication by exactly $\frac{3}{2} = 1.5$, but is now multiplication by

$$2^{1/12} \times 2^{1/12} \times 2^{1/12} \times 2^{1/12} \times 2^{1/12} \times 2^{1/12} \times 2^{1/12} = 2^{7/12} \sim 1.498,$$

that is, going up seven semitones in the equally tempered scale
is pretty close, but not exactly the same as going up a perfect
fifth. Similarly for perfect fourths and the other tones in the
Pythagorean scale, though the octave remains the same. But
musically, these are outweighed by important advantages. We
can now change keys freely and often, with all of the intervals
sounding the same from one key to the next. We give up on
perfection, but we gain a huge amount musically.

The 12 notes in the equally tempered scale (often called the
chromatic scale, are more than the 7 notes in the Pythagorean (major) scale—it inserts some notes between those of the Pythagorean
scale in order to have equally spaced notes (according to ratios).
The notes in the equally tempered scale that correspond most
closely to those of the Pythagorean scale are $0, 2, 4, 5, 7, 9$ and

Note that when we raise powers to powers, we multiply
the powers, so that raising
both sides of $r^{12} = 2$ to the
power $1/12$, we get:

$$(r^{12})^{1/12} = 2^{1/12}$$
$$r^{12/12} = 2^{1/12}$$
$$r = 2^{1/12}$$

11 semitones up from the starting note (see Table 7.2). In the equally tempered scale, the frequencies are slightly different from those of the Pythagorean scale. For example, starting with a middle C with a frequency of 262, as before, we get:

Note	Pythagorean frequency	Equally tempered frequency
C	262 Hz	262 Hz
C♯		278 Hz
D	295 Hz	294 Hz
D♯		312 Hz
E	332 Hz	330 Hz
F	349 Hz	350 Hz
F♯		371 Hz
G	393 Hz	393 Hz
G♯		416 Hz
A	442 Hz	441 Hz
A♯		467 Hz
B	497 Hz	495 Hz
C	524 Hz	524 Hz

Table 7.2: Comparison of frequencies for the notes in the Pythagorean and equally tempered scales starting at middle C. There are extensions of the Pythagorean scale that have 12 notes with frequencies close to those of the equally tempered scale, but we won't pursue that here, as their frequencies are not (and for mathematical reasons, cannot be) universally accepted.

You might ask why it was decided to divide the octave up into 12 equal steps rather than some other number of steps. The answer is again mathematical, as 12 is the smallest number of steps that would allow for very good approximations to the seven notes in the Pythagorean scale (as Table 7.2 can attest).

In terms of musical notation, the same staff system is used, with the major equally tempered scale shown as before (even though the frequencies of the notes differ somewhat from the Pythagorean major scale). For notating the extra notes beyond the major scale, one adapts the staff notation by using the *sharp* symbol, ♯, to indicate the movement of a note up one semitone (or, equivalently, the *flat* symbol, ♭, for movement *down* one semitone).

Understanding the role that the twelfth root of 2 plays in music is vital in building instruments like guitars. If you were to build a guitar (I had a friend in high school who did just that) how would you figure out where to place the frets? It's known

The major scale is defined by the sequence of 2, 2, 1, 2, 2, 2, 1 semitones that forms the scale, starting at a base note. There are other common scales, including the *natural minor*, with sequence 2, 1, 2, 2, 1, 2, 2 of semitones, the *harmonic minor* scale, used in eastern European music, with sequence 2, 1, 2, 2, 1, 3, 1, and the *pentatonic* scale, the basis of blues and oriental music, with sequence 3, 2, 2, 3, 2.

C C# D D# E F F# G G# A A# B C

Figure 7.6: Chromatic scale starting at middle C, C4.

that the frequency that a plucked string resonates at is inversely proportional to the string's length, so that if you double the length of the string, you cut the frequency it resonates at in half, and if you cut the length into one-third its original length, the frequency becomes three times as large, and so on. Mathematically speaking, if f is the frequency that a string resonates and l is the length of the string, then for some fixed number K that doesn't change (but depends on the type of string and its tension) we have

$$f \times l = K,$$

that is, the frequency and string length are **inversely proportional**.

We can use this relationship between the frequency and the string length to locate all of the frets on a guitar neck. For example, for one of my guitars (see Figure 7.7), I measured the length of the strings from the nut at the top to the bridge at the bottom (these are the free parts of the strings that vibrate when plucked) to be 28 inches. Let the frequency at which one of the strings resonates when plucked be f (we'll find out that we don't even need to know what this is!). The first fret should be placed where the shortened length of the string (from the fret to the bridge) will move the frequency up a semitone from f; we'll call this length l. Moving up a semitone corresponds to multiplication by $2^{1/12}$, so the new frequency will be $2^{1/12} \times f$. Using the fact that the frequency times the string's length always stays the same, we find that $f \times 28 = (2^{1/12} \times f) \times l$. We can cancel out the f on both sides and divide through by $2^{1/12}$ to find that l, the length of the string from the first fret down to the bridge, must be $l = 28/2^{1/12}$, which a calculator will show is about 26.4 inches. So the first fret should be $28 - 26.4 = 1.6$ inches from the nut at the top of the guitar (and this is precisely where it is on my guitar).

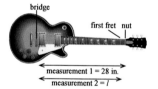

bridge

first fret nut

measurement 1 = 28 in.
measurement 2 = l

Figure 7.7: Nut and bridge of a guitar.

Exercises

Exercise 7.1.1. For the following sine curves, state the frequency and the amplitude:

(a) $2\sin(2\pi \times 440 \times t)$ (b) $1.2\sin(2\pi \times 262 \times t)$ (c) $\sin(2\pi \times 100 \times t)$

(d) $0.7\sin(10\pi t)$ (e) $1.1\sin(440t)$

Exercise 7.1.2. For the following frequencies and amplitudes, write out the formula for a corresponding sine curve:

(a) frequency 440 Hz, amplitude 2.5   (b) frequency 220 Hz, amplitude 0.75

(c) frequency 267 Hz, amplitude 1

Exercise 7.1.3. A kilohertz (kHz) is 1000 hertz. This yields conversion factors (as discussed in Section 1.2) of $\dfrac{1000\ \text{Hz}}{1\ \text{kHz}}$ and $\dfrac{1\ \text{kHz}}{1000\ \text{Hz}}$. For each of the following, convert the frequency in hertz to kilohertz.

(a) 1000 Hz   (b) 1500 Hz   (c) 220 Hz   (d) 5 Hz

Exercise 7.1.4. For each of the following, convert the frequency in kilohertz to hertz.

(a) 1 kHz   (b) 32 kHz   (c) 0.75 kHz   (d) 0.01 kHz

Which are within the range of human hearing?

Exercise 7.1.5. Convert the list of frequency ratios of the Pythagorean scale to decimals and compare them to the decimals for the corresponding ratios for the equally tempered scale.

Exercise 7.1.6. Write out the frequency of the Pythagorean A major scale starting on A 220 Hz.

Exercise 7.1.7. The ratio between a C and D in the Pythagorean scale is 9/8. This is a "whole tone" in the Pythagorean scale, and six whole tones should take you to an octave, which has a frequency ratio of 2 (that is, the octave has twice the frequency of the original starting note). If you go up six whole tones, starting at C, at what frequency ratio do you end up? How does this compare to the frequency ratio of the octave?

Exercise 7.1.8. Repeat Exercise 7.1.7 with 12 semitones in a an octave, taking the semitone between the third and fourth notes (or the seventh and eighth notes) in the Pythagorean scale.

Exercise 7.1.9. If someone plays the two tones at 60 Hz and 90 Hz, what combination tone do you think might be heard?

Exercise 7.1.10. Starting with a frequency of A 220 Hz, find the frequency of the following notes in the equally tempered tuning:

(a) the whole tone B above A   (b) the whole tone G below A   (c) the G♯ above A

Exercise 7.1.11. For each of the following notes, find its frequency (it may be helpful to compare it to A 220 Hz):

(a) G above middle C   (b) C below middle C   (c) F♯ an octave and a half above middle C

Exercise 7.1.12. For each of the following frequencies, find the corresponding note:

(a) 440 Hz   (b) 262 Hz   (c) 165 Hz   (d) 880 Hz   (e) 353 Hz

Exercise 7.1.13. Write out the equation of a wave that would correspond to a note at a frequency of 220 Hz along with another note that is a perfect fifth up, at (approximately) half the volume (do this for both the Pythagorean and equally tempered scales).

Exercise 7.1.14. Show that going up a perfect fourth is the same as going down a perfect fifth (again, in both the Pythagorean and equally tempered scales).

Exercise 7.1.15. In the Pythagorean scale, will some stacking of perfect fourths take you back to the same pitch class you started in? Why or why not?

Exercise 7.1.16. Using the program "Trig Sounds" at www.jasonibrown.com/mla, construct and listen to the beats that occur when two frequencies are close together. Using a stopwatch, count how many beats you hear in a second compared to the difference between the frequencies. What do you find?

Exercise 7.1.17. Using the program "Trig Sounds" at www.jasonibrown.com/mla, start with a base tone of 100 Hz and try adding in the first harmonics (that is, whole number multiples of the original frequency: 200 Hz, 300 Hz, 400 Hz, and so on), varying the amplitudes. How would you characterize the sounds, based on the relative loudness of the harmonics?

Exercise 7.1.18. Take a guitar and a tape measure. Measure the length of the strings from the nut at the top of the guitar (where the strings pass over and travel down the neck) and the bridge at the bottom. Then measure the length of the strings from the first fret down to the bridge. Divide the first number by the second with a calculator. What is this about?

Exercise 7.1.19. Complete the measurements of the first 12 frets on a guitar with a string length of 28 inches.

Exercise 7.1.20. Make a table of the ratios for the intervals from unison until the octave for both the Pythagorean and equally tempered major scales, using decimals. For which notes in the C major Pythagorean and equally tempered scales is the absolute value of the difference between the ratios of the frequencies the biggest? Can you hear the difference? Try using the program "Trig Sounds" at www.jasonibrown.com/mla to check this out.

Exercise 7.1.21. Using the program "Build Your Own Scale" at www.jasonibrown.com/mla, try building an equally tempered scale with more or less than 12 notes (choose the "geometrically" radio button), and test out the keyboard. If you have chosen 15 notes for your scale, what is the ratio between successive notes?

Exercise 7.1.22. You can use the program "Build Your Own Scale" to try out a scale where you set the next note by equally spacing (arithmetically) the notes out between a note and its octave. What do you think of the sound?

Exercise 7.1.23. (Harder) Try out some other numbers of semitones (besides 12) to divide up an octave in equal temperament, that is, with equal ratios between steps. Can you find another number of semitones that approximates the Pythagorean small fraction ratios even better than the 12 semitones do?

Exercise 7.1.24. If you divide up the interval of a semitone into 100 equal ratios, each is called a *cent*. What is the decimal corresponding to a cent?

Exercise 7.1.25. A *chord* consists of at least three notes, played simultaneously. A *major chord* consists of a major third and a minor third stacked in that order, while a *minor chord* consists of the same two intervals, stacked in the opposite order. Three note chords are traditionally referenced by roman numbers, upper case for major chords, lower case for minor chords. So, for example, the major chord on the first note of the major scale is denoted by I, and starting on C, it is the chord C, E, G, while the minor chord starting on the second note of the major scale is denoted by ii, and for the C major scale, is the collection of notes D, F, A.

What stacking of two of the same intervals will yield the same chord (i.e., the same three pitch classes) no matter which note of the chord you start on? Find the name of such a chord (it is not a major or minor chord).

Exercise 7.1.26. An interval consists of two notes. The table below describes the names of the intervals, based on the number of semitones up between the lower note and the higher one.

Name	Number of Semitones
unison	0
minor second	1
major second	2
minor third	3
major third	4
perfect fourth	5
tritone	6
perfect fifth	7
minor sixth	8
major sixth	9
minor seventh	10
major seventh	11
octave	12

Suppose you stack two intervals and get a perfect fifth. What intervals did you use?

Exercise 7.1.27. In the equally tempered scale, how many perfect fifths do you have to stack to get back into the same pitch class?

Exercise 7.1.28. In the equally tempered scale, how many minor thirds do you have to stack to get back into the same pitch class?

Exercise 7.1.29. Characterize mathematically the intervals for which stacking less than 12 of them gets you back into the same pitch class.

Exercise 7.1.30. Explain why, if you notate a given note on a musical staff, then if it is marked on a space, its octave up or down is marked on a line, and vice versa.

Exercise 7.1.31. The *just* scale is similar to the Pythagorean scale, but with slightly different small ratio intervals:

$$1, \frac{9}{8}, \frac{5}{4}, \frac{4}{3}, \frac{3}{2}, \frac{5}{3}, \frac{15}{8}, 2.$$

(It can even be said that these ratios are smaller than those of the Pythagorean scale, so they better fit the concept of a "small ratio" scale.) Which intervals are different between the Just and Pythagorean scales? Compare the difference as decimals.

7.2   *Mathematics of Composing*

We have seen the integral part that mathematics plays in our listening to music—from trigonometry functions in our perception of pitch and loudness, to the use of small fractions and powers to describe notes and scales. Now, of course, the music we listen to needs to be composed. And once again, mathematics plays a crucial role in the aesthetics of music. You'll find that what often moves you emotionally music is the underlying mathematics.

How Much Music Is Left to Be Written?

One of the greatest writer's blocks for musicians is the feeling that they have heard or written everything. Haven't we been exposed to every sequence of notes, rhythms and harmonies? Some basic mathematics shows how far this is from the case. We'll start with melodies.

A *melody* is a sequence of notes that are sung or played by an instrument. These are the tunes we go around humming, if they are catchy. A simple melody might be

E4, D4, C4, E4, D4, C4, G4, F4, F4, E4, G4, F4, F4, E4,

which is the melody for the beginning of the nursery rhyme "Three Blind Mice." Of course, this is a simple song, and the melody reflects that. We can see obvious repetitions, which is an elementary mathematical component of music composition, but more about that later.

Any sequence of notes gives rise to a melody, though some would be more interesting and appealing than others. But aesthetics is in the eye, ear and mind of the beholder; some melodies that one person would find difficult and hard to listen to (often called "dissonant") would be interesting and appealing to another. So let's, for the time being, ignore the question of whether a melody sounds "good" or not and ask how much variety we have when we sit down to write a melody. Certainly we have heard a lot of good (and a lot of bad) melodies. Is there, as King Solomon said, nothing new under the sun?

To help decide, we rely on our counting principles from Chapter 4. Suppose we were interested in how many melodies there were with say 10 notes. In music, it usually doesn't matter on which note you start, all that matters is where you go from there. So let's consider how many 10-note melodies we can write starting on middle C. Theoretically, we can go up or down any number of notes, but that would be hard to sing (or even play). Let's suppose we restrict ourselves to going up or down at most an octave, that is, at most 12 notes (or semitones). So from any given note, we have 25 choices for the next note in the sequence (any one of the 12 notes up or down, plus possibly staying on the same note):

C4 $\underline{}$ $\underline{}$ $\underline{}$ $\underline{}$ $\underline{}$ $\underline{}$ $\underline{}$ $\underline{}$ $\underline{}$

As we fill each successive blank there are 25 choices, so we use the multiplicative principles to find out that there are

$$25 \times 25 \times 25 \times 25 \times 25 \times 25 \times 25 \times 25 \times 25 \times 25$$
$$= 25^9$$
$$= 3,814,697,265,625$$

many such melodies, that is, about 4 *trillion* of them! And these don't even count melodies that go beyond an octave up or down from the starting note, nor do they take into account any rhythm for the melody. Have you heard all of these? I think not! So there is, indeed, much musically new under the sun.

Feel the Rhythm of the Beat

Rhythm is inherently mathematical. Just as we divide up
an octave into two equally spaced frequencies (the notes of the
chromatic scale), we can divide up time into variously equal
intervals. We can feel and enjoy the periodic pulses from doing
so, and the various patterns we can derive by repeating rhythms
can drive us to sit up and listen or even get on our feet and
dance. It's the mathematics in the beats that makes us go gaga
(even if we aren't Lady Gaga).

For example, in a lot of western music (like rock 'n' roll) a
musical time interval (called a *bar* or *measure*, is often divided up
into four equal parts (see Figure 7.8).

Figure 7.8: Musical bar, and
the bar divided into four.

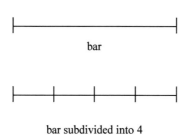

bar

bar subdivided into 4

If we strike a drum, strum a guitar or pluck a bass at the
beginning of each interval (called a *beat*), we feel a rhythm, a
pattern (the > symbol below shows each beat's start, often called
an *onset*). For many popular songs, you can feel the bass note
played on each of these four beats. The more the pattern repeats,
the more ingrained it is in our minds.

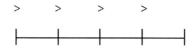

If we let the duration of a whole bar be "denoted" by a *whole
note*, then the duration of each of the four equal notes, which
last a quarter of the entire bar, is a quarter of the whole; they are

each called *quarter notes*. (*Half notes* are half as long as a whole note, and take up the time of two quarter notes.)

Now this process of subdivision can be carried out further. Each of the quarter notes above can be divided into smaller, equal size subintervals. Most often, the subdivisions are into two equal parts (often in classic rock) and four equal parts (modern rock, hip hop, rap and disco), though occasionally into three equal parts (slow blues, shuffle, swing). Musicians, when they write music, use a line at the top of a quarter note to indicate an *eighth* note (which lasts half as long as a quarter note) and two lines at the top to indicate a sixteenth note (which is half the duration of an eighth note, a quarter the duration of a quarter note, and a sixteenth the duration of a whole note).

Music writers also often group these together by joining tops for readability. Here is what it would look like to subdivide each quarter note into four *sixteenth* notes. You can continue to subdivide each note as needed.

Of course, seeing the mathematics suggests that we needn't restrict ourselves to subdividing each note up into two, three or four equal parts. How about five or more? It certainly makes sense. You'll find very few popular songs that do so. Why? It seems we humans have an innate ability to divide things up into a small number of equal parts—into halves easily, into thirds a little less easily, and into fourths about the same, though we can simplify the latter by dividing time in half and then half again. Dividing time into equal fifths and sixths is difficult for most people, whether you are a musician or a listener. A more complicated rhythm involves collections of these patterns (and rests, or silences, of similar durations), but in each full bar the sum of the total durations must be equal to that of a whole note. To create a rhythm, you can freely choose where to strike or strum. For example, if the smallest duration is to be a sixteenth

Figure 7.9: Whole note (lasts the entire bar).

Figure 7.10: Quarter notes (each lasting 1/4 of the entire bar).

quarter note

eighth note

sixteenth note

Figure 7.11: Musical notation for notes of various durations.

note, then, out of the 16 sixteenth notes in a bar, we can choose to strum our guitar on the first, second, third, fourth, fifth, ninth, eleventh, twelfth, fourteenth and sixteenth notes. The $>$ signs indicate where to strum (see Figure 7.12).

Figure 7.12: Rhythm with sixteenths.

Repeating a rhythm is catchy. So now we ask: how many *grooves* (i.e., repetitive rhythmic patterns) can we make? We'll only consider grooves of the length of one bar, and since on many percussive (or rhythmic) instruments like drums we can only hear onsets rather than durations of notes, we'll only consider where the onsets (that is, the strikes) are. If we only allow ourselves to use quarter notes, then we can have a choice for each of the four quarters—to strike or not to strike, that is the question!

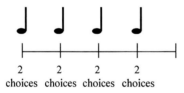

Counting such choices is again an application of the multiplicative principle. For each of the four quarter notes, we have two choices—strike or not to strike. The multiplicative principle states that the total number of ways to make all choices is $2 \times 2 \times 2 \times 2 = 2^4 = 16$. Therefore there are only 16 such grooves. Not so many.

If we allow eighth notes, then we have a bar where we can strike or not strike on each of the eight eighth notes. Counting the same way, we find that there are

$$2 \times 2 \times 2 \times 2 \times 2 \times 2 \times 2 \times 2 = 2^8 = 256$$

groves with eighth notes, a lot more.

What about if we go to sixteenth notes? Again counting using the multiplicative principle, we find that there are $2^{16} = 65\,536$ grooves—a whole *lot more*! So once again, mathematics comes to show us how vast our musical choices are.

Moreover, the math gives us a systematic way to try to list all of our choices if we wanted to. To do so, we can think of making a *tree*, where the branches correspond to choices. Let's go back to looking at grooves using just quarter notes. Our first choice is decide whether to strike the first or not, so we indicate this with a tree that has two branches, strike or don't strike:

strike

don't strike

For each of these choices, we still have two more choices to make—strike or not strike the second quarter note.

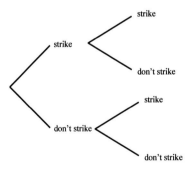

strike

strike

don't strike

don't strike

strike

don't strike

We continue on for the last two notes, getting the tree shown in Figure 7.13.

The tree encodes our choices, and each complete branch from the leftmost point and ending at a rightmost point corresponds to a complete set of choices. For example, the top branch indicates that we strike on each of the four quarter notes, the third that we strike on the first, second and fourth quarter notes but

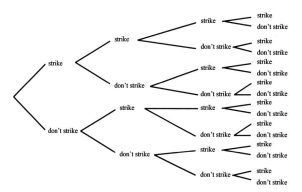

Figure 7.13: A decision tree for rhythm.

not on the third, and so on. The number of branches, 16, is the total number of possibilities (as we have seen before), and each branch lists a different possibility (or groove, in our case). Here is a listing of all 16 grooves with 4 quarter notes (the squiggly symbol, if you haven't seen it before, is a *quarter rest*) and corresponds to <u>not</u> playing on the associated possible onset; the first column corresponds to the top eight branches of the tree, the second column to the remainder:

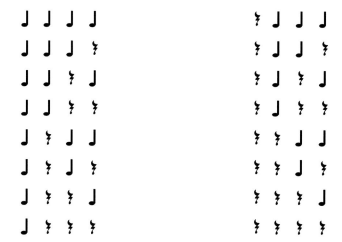

You can use exactly the same method to list all rhythmic grooves on eighth notes or sixteenth notes, but the tree of course gets very large.

Now let's try counting a more difficult problem—the number of rhythmic grooves where we decide ahead of time how many onsets (or strikes) we are going to have. Let's focus on a particular example, say 7 onsets among a bar of 16 sixteenth notes. This question is the same as deciding how many ways can we choose 7 numbers (the onsets) from among 16 different possibilities (the 16 different numbers 1 through 16 inclusive).

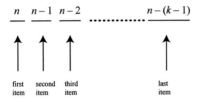

This sounds suspiciously like combinations, and indeed it is. The number of ways to select 7 onsets among a bar of 16 sixteenth notes is $C_{16,7}$, which is

$$\frac{16!}{(16-7)!7!} = \frac{16!}{9!7!} = 11\,440$$

many such grooves, and, more generally, in a bar with n possible onsets, the number of grooves with k onsets is exactly $C_{n,k}$.

Now here are a few mathematical rules about combinations that we can see just from the connection to counting grooves with k onsets among n possible ones. For example, choosing the k onsets is obviously equivalent to choosing the ones we miss (of which there are $n - k$), so reversing the roles of choosing onsets and non-onsets, we see that the number of grooves with k onsets among n possible ones is the same as the number of grooves with $n - k$ onsets among n possible ones, that is,

$$C_{n,k} = C_{n,n-k}.$$

Also, we can consider grooves with k onsets among n possible ones by breaking them down into two cases—those for which there is an onset on the first possible onset, and those for which there isn't. If we choose the first possible onset as an onset, then we choose the remaining $k-1$ onsets from among the $n-1$ possible onsets remaining (all but the first); there are $C_{n-1,k-1}$ of these. If we don't choose the first possible onset, then we choose the remaining k onsets from among the possible onsets remaining (all but the first); there are $C_{n-1,k}$ of these. But we know there are many ways to select k onsets from n possible ones, so we see from the additive principle that

$$C_{n,k} = C_{n-1,k-1} + C_{n-1,k}.$$

This is what is known as a **recurrence relation**; the value of some function is given in terms of the values of "smaller" instances of the function and is typical of a general mathematical principle of reducing a problem to some previously solved (or found) ones. For example, suppose we know $C_{n,k}$ for all values of n smaller than 4 and all values of k (being between 0 and n, of course):

In the previous chapter you were introduced to the best known recurrence relation, the **Fibonacci sequence** $1, 1, 2, 3, 5, 8, 13, \ldots$

$$C_{1,0} = 1 \quad C_{1,1} = 1$$
$$C_{2,0} = 1 \quad C_{2,1} = 2 \quad C_{2,2} = 1$$
$$C_{3,0} = 1 \quad C_{3,1} = 3 \quad C_{3,2} = 3 \quad C_{3,3} = 1$$

Then to calculate $C_{4,2}$, we use the recurrence relation to write $C_{4,2} = C_{3,1} + C_{3,2}$. But we know that $C_{3,1} = 3$ and $C_{3,2} = 3$, so we find that $C_{4,2} = 3 + 3 = 6$. We can theoretically calculate *any* value of $C_{i,j}$ from this recurrence.

And the recurrence relation tells us even more. For we derived $C_{n,k} = C_{n-1,k-1} + C_{n-1,k}$ by the additive principle of counting, where we said we broke up the list of rhythms with k strokes among n possible onsets by either

- striking on the first possible onset, followed by any rhythm of $k-1$ strokes out of the $n-1$ possible onsets left, or

- missing the first possible onset, followed by any rhythm of k strokes out of the $n-1$ possible onsets left.

The counting provides us with what is called a **recursive method** (that is, building new items from previous ones) for *listing* rhythms. So, for example, Figure 7.14 is a list of all ways to make two strokes out of four possible onsets and Figure 7.15 is a list of all ways to make three strokes out of four possible onsets.

Figure 7.14: All ways to make two strokes out of four possible onsets.

Figure 7.15: All ways to make three strokes out of four possible onsets.

We can form a list of three strokes out of five possible onsets by attaching a stroke to the front of each one in the first list, and a rest to the front of each item in the second list (see Figure 7.16).

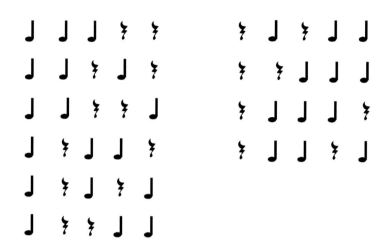

This gives us a total of 10 such rhythmic grooves.

Now you might think to yourself, "I'm only interested in rock music, with four beats in a bar. Why should I even consider bars with five beats in them?" Well, the story you learn here is that even if you want to count (or list) rhythms for bars consisting of four quarter notes, you might need to consider bars of other lengths in order to build up to the ones in question (we might be ultimately interested in a bar divided into eight possible onsets). This is a lesson from mathematics you can carry on to all of your problem solving:

Life Lesson: Sometimes to solve a particular problem, it will help to solve other related problems, even if you aren't interested in the latter.

Patterns, Functions and the Art of Writing Great Music

We like patterns. We just can't help it; it's part of being human. But we also appreciate surprises. More specifically:

Life Lesson: Great music depends on both setting up expectations and intentionally breaking them. And expectations depend critically on mathematics—the listener will have to do the mental calculations to know what to expect.

The expectations can be set and broken, simultaneously or independently, in the melody, the harmony or the rhythm, or even the lyrics. Those who can find the perfect mix between setting and breaking expectations are destined for longevity and greatness, whether it is in art, science or music.

Let's look at some simple musical examples. Rhythmic grooves depend on repetition to become ear worms and get inside your head. If the pattern is really interesting and repetitive, you'll have a tough time shaking it out of your brain. A simple pattern that made many a successful disco song is to have the bass and the bass drum play repetitive quarter notes over and over again:

This rhythm is regular, like a heartbeat, so you'll feel it in your chest and feel you'll want to get up on the dance floor. (The three dots at the end indicate to repeat ad nauseum). You can, of course, build more interesting patterns to repeat. Here is the rhythm of the riff that The Beatles play on "Day Tripper"; the pattern uses eighth notes as a base, and I indicate the actual onsets as usual with the > sign:

Listen to The Beatles' "Day Tripper," and try to count out the rhythm of the riff. Isn't the rhythm a big teaser?

If you listen to it, you'll find it quite catchy, I think.

Rock 'n' roll is based on bars of 4 quarter notes, subdivided into 8 eighth notes (and sometimes into 16 sixteenths). The natural groupings are into groups of four (quarters, sixteenths) and groups of two (eighths). What happens when you choose to group a rhythm differently on top of these underlying rhythms? You get something so catchy that it can make you walk like a duck!

Let me explain. If we take a bunch of eighth notes and create patterns of three eighth notes against the natural background of four eighth notes (each bar contains two pairs, a strong beat—the first and third quarter beat—and the following "backbeat"— every second and fourth quarter beat), then we get a wild feeling to the music. It's not so easy to do at first—it takes practice just like it might to pat your head while rubbing your stomach. But the practice is well worth the end result (at least for the music).

How does the sound affect the audience? Well, the listener waits for the rhythm on top, in groups of threes, to match up with the underlying groups of fours—the expectation is naturally set up that the two patterns (the foreground pattern of threes against a background pattern of fours) will eventually align. When will it do so? We want some number of threes to be equal to some number of fours, and the first time this happens is at the **least common multiple (LCM)** of three and four, which happens to be 12; 4 threes and 3 fours are both 12. And this is what you often hear with patterns of threes against patterns of fours—four patterns of threes played, at which point the player exits the pattern and plays something else. If it is done right, with a good "exit strategy" as described, everyone will be happy—player and listener alike. Figure 7.17 is an example; note the groupings of three eighth notes (a quarter and an eighth) four times until it breaks out into other notes:

The least common multiple of two positive integers a and b is the smallest positive integer that both a and b divide. So, for example, the least common multiple of 18 and 30 is 6, as 6 is the largest positive integer that divides both.

x x x x x x x x | x x x x x x x x
1 & 2 & 3 & 4 &   1 & 2 & 3   & 4 &

Figure 7.17: A Chuck Berry-like example of groupings of 3s against background groupings of 4s.

In fact, this is exactly what Chuck Berry, the grand old man of rock 'n' roll lead guitar, does in many of his guitar solos. And he finds the math-in-the-music so moving that he can't help but do his famous *duck walk* at such a time. It's quite spectacular what a little math can do.

You'll find other such examples throughout pop music. George Harrison loved this trick; see if you can spot it in his guitar work on "A Hard Day's Night" and "Here Comes the Sun." Roger McGuinn, the lead guitarist for the group The Byrds, also enjoyed a least common multiple or two—have a listen to his guitar riffs in the songs "Mr. Tambourine Man" and "Turn, Turn, Turn."

And the overlapping of patterns of different lengths needn't be only in the guitarist's domain. What makes Led Zeppelin's song "Rock and Roll" so unusual is the patterns of three eighth notes against the background of twos and fours ("been a long time, been a long time, been a long time" and "lonely, lonely, lonely, lonely" are each examples).

Now the one thing that begs to be asked is what about other groupings on top of one another? The threes against fours is the most common one I've found, but there are certainly examples of others. What the mathematical viewpoint suggests is trying out different length patterns against one another, even if they are difficult or unnatural to play at first. This is the key to reaching out to new ideas in music—moving beyond what you are comfortable playing and listening to. And the math here is a perfect example of that.

> *Life Lesson:* Sometimes mathematics suggests alternatives you might not have thought of otherwise. Explore the possibilities.

Let's look at another technique for creating and modifying interesting patterns. In math we're used to taking objects and applying functions or **transformations** to them. For example, if we take the number 4 and apply the function $f(x) = x + 5$, we get $f(5) = 4 + 5 = 9$. Of course, we can apply the function to each number in a sequence of numbers, say

$$0, 0, 3, 4, 7, 7, 9, 7$$

to get

$$5, 5, 8, 9, 12, 12, 14, 12.$$

The patterns are of course different, but somehow the same.

How do functions like these interact with music? In a lot of different ways, because we can apply simple functions to any musical patterns, whether they be in the melody, the harmony or the rhythm, and in general we'll find the whole result interesting yet satisfying. For example, in many rock and blues songs, the players often construct a *riff*, that is, a short sequence of notes (or chords), and then move each note up or down (i.e., *transpose* it in music terminology) a certain number of semitones. The result is very catchy indeed. So if we take a riff like

$$A, A, C, C\sharp, E, E, F\sharp, E$$

and move it up five semitones (that is, a perfect fourth), then we get

$$D, D, F, F\sharp, A, A, B, A,$$

which sounds much like the original riff, but has some movement and interest to it. If we start off calling A note "0" and write down how many semitones each note is above A, then the riff

$$A, A, C, C\sharp, E, E, F\sharp, E$$

is precisely $0, 0, 3, 4, 7, 7, 9, 7$, the riff $D, D, F, F\sharp, A, A, B, A$ is $5, 5, 8, 9, 12, 12, 14, 12$, and we are applying the function $f(x) = x + 5$ to the sequences. It's really just math, but it sounds like so much more.

What about if we apply such a simple function to the rhythm? The effect can even be more astounding. Listen to the opening of The Beatles' "I Want to Hold Your Hand." The rhythm grabs hold of you until the song enters, each time you hear it. Why is it so good? Well, listeners have a natural subconscious yearning to determine beat one (where each bar starts) to orient themselves to the rhythm.

So a musician can play a good "joke" on the audience by playing around with this natural assumption and "hiding" beat one. Each time we listen to the opening of "I Want To Hold Your

Riffs are an essential component of rock music. You'll often hear a guitar and bass play the riff in unison. Beatles songs like "I Feel Fine," "Drive My Car," "Day Tripper" and "Birthday" all are based on riffs that are incredibly catchy, as they set expectations and undergo mathematical transformations. Have a listen!

Figure 7.18: "If you listen to the great Beatles records, the earliest ones, where the lyrics are incredibly simple. Why are they still beautiful? Well, they're beautifully sung, beautifully played, and the mathematics in them is elegant." (Bruce Springsteen, 2009)

Hand," we experience it as on the left, with the first strum being on beat one (above the notes I drew in *chord diagram* for those guitarists out there who want to play along):

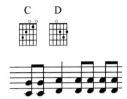

But those tricksters didn't play this; they applied the mathematical *transformation* $f(x) = x - 3$ and moved the rhythm back three eighth notes:

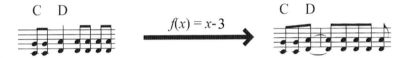

What this does is confuse the listener when the melody comes in—you can't easily find beat one. You are left feeling the next time you listen to the song as if you can't quite figure out the beginning. The transformation makes the opening of the song eternally ambiguous, hiding beat one—and in the arts, unlike real-life, ambiguity is a *good* thing!

Of course, you aren't limited by functions only of the type $f(x) = x + a$, where a is some integer (positive a means to move up by that number of semitones, while negative a means to move down); such functions are called **translations**. You can try out all sorts of things, like **compressions** and **stretches** of the form $f(x) = ax$, **reflections** of the form $f(x) = -x$, and various combinations of all of these. You don't need to think of the numbers or the functions, but realize that if you are doing the same thing to all notes, a transformation lies hidden.

As another case in point, listen to the song "Hot Blooded" by the rock group Foreigner. What Foreigner does with the pattern of chords is pure mathematical (and musical) genius. The guitar riff (and indeed the song) is built around a three chord pattern, with the first repeated, played behind the title

Figure 7.19: The band Foreigner.

words (for guitarists, the chords are Esus4, Esus4, E). First, in the next two bars they move the pattern up five semitones (corresponding to applying the function $f(x) = x + 5$) while moving it back in time two eighth notes (corresponding to applying the function $f(x) = x - 2$ to the rhythm) and reflecting the top notes in the middle line (corresponding to applying the function $f(x) = -x$), repeating these two bars twice more (that is, applying the function $f(x) = x$, which doesn't move anything, to the pattern of the first two bars), and finally, in bar 7, moving the pattern up seven semitones (corresponding to applying the function $f(x) = x + 7$) and repeating this bar, moving the pattern back two eighths in time (function $f(x) = x - 2$ again).

The net result is that the simple one-bar pattern is converted to a more complex eight-bar pattern that holds together, as it is all based on the first bar, but introduces interest by the various elementary transformations applied to different aspects (pitch, rhythm and to combined segments of the pattern).

What do we learn from of all of this? We find that even some very simple music can grab your attention indefinitely if you transform it with some simple functions. And your listener may never even know that it's the math that they really like!

Life Lesson: Transformations on music—in the melody, the harmony, the rhythm, the riffs, and even the lyrics—allow the listener to compress the music while keeping interest, and that is ideal!

Exercises

Exercise 7.2.1. The melody of the song "Row, Row, Row Your Boat," starting on middle C, is
        C4,C4,C4,D4,E4,E4,D4,E4,F4,G4,C5,C5,G4,G4,E4,E4,C4,C4,G4,F4,E4,D4,C4.
If we denote middle C by 0, write out the sequence in terms of the number of semitones each note is up from middle C.

Exercise 7.2.2. Transpose the melody of "Row, Row, Row Your Boat" up a whole tone (that is, two semitones) by first adding 2 to each number in the corresponding sequence (see the previous problem) and then transferring the sequence back to notes.

Exercise 7.2.3. Count melodies of 10 notes, starting on middle C, that lie between C3 and C5 inclusive. How does this compare to the example done in this section, where we start on middle C, but can only go up or down at most an octave?

Exercise 7.2.4. The basic chords of the major scale are the major chords I, IV and V, while the lesser used chords are ii, iii and vi. How many sequences of seven chords can we make just using the major chords, with the rule that no chord can be the same as the previous one? How many such sequences with all six chords?

Exercise 7.2.5. (Harder) How many sequences of five chords can we make with just the major chords I, IV and V, with the rule that no chord can be the same as the previous one and the first and last are different as well (so that you can repeat the sequence over and over)?

Exercise 7.2.6. List all sequences satisfying Exercise 7.2.5.

Exercise 7.2.7. A guitarist's fingers span the first four frets on a guitar. On each string, a finger (or thumb) can either be placed just behind one of these frets, or the string can be left open. Some fingers can be "barred," that is, can press down on more than one string. Show that the number of different sounds that the guitarist can play by a single strum after putting his fingers down is *at most* 15 625.

Exercise 7.2.8. For Exercise 7.2.7, suppose that we insist that the guitarist cannot barre nor use his thumb. Show that the number of different sounds is at most 5835.

Exercise 7.2.9. How many quarter notes are equal to three whole notes?

Exercise 7.2.10. How many sixteenth notes are there in four bars of three quarter notes each?

Exercise 7.2.11. A *thirty-second* note has a duration that is half of a sixteenth note. In a bar of four quarter notes, how many thirty-second notes are there?

Exercise 7.2.12. The time signature is a fraction of numbers (written without the horizontal bar), where the denominator indicates what note duration is the basic one, and the top indicates how many of those are in the bar. So, for example,

- the basis of rock music, $\frac{4}{4}$ (spoken as "four four time"), indicates that a quarter note is the basic unit in the bar, and a bar consists of four quarter notes,

- waltz time ("three four time") common in European and country music, is $\frac{3}{4}$, and

- $\frac{12}{8}$ ("twelve eight time") indicates that an eighth note is the basic unit in the bar, and a bar consists of 12 eighth notes (this time signature is common in slow blues and shuffles).

Explain how $\frac{4}{4}$ and $\frac{2}{2}$ are essentially the same.

Exercise 7.2.13. A *triplet* is a grouping of three eighth notes. How many ways can you make a groove on a bar of four triplets (that is, a bar of twelve eight time)?

Exercise 7.2.14. List all the ways you can subdivide a bar of $\frac{12}{8}$ time into groups of the same number of eighth notes? For example, a bar of $\frac{12}{8}$ time can be broken up into four triplets (groups of three eighth notes).

Exercise 7.2.15. How many ways are there to choose three onsets from a bar of eight possible onsets, if we can't strike all three in a row?

Exercise 7.2.16. List the grooves for Exercise 7.2.13.

Exercise 7.2.17. How many grooves of 5 onsets, chosen from 16 possible ones, are there when you insist the first onset is on the first possible one?

Exercise 7.2.18. How many grooves of 5 onsets, chosen from 16 possible ones, are there when you insist the first onset is after the first possible one?

Exercise 7.2.19. Explain why there are as many grooves of a bar of 16 possible onsets (on the sixteenths) as there are grooves of two bars of 8 possible onsets (on the eighths) in each?

Exercise 7.2.20. Use grooves to explain why $C_{n,k} = C_{n-1,k-1} + C_{n-2,k-1} + \cdots + C_{k-1,k-1}$ holds for any values of n and k where $k \leq n$. (Hint: Consider where the first onset occurs.)

Exercise 7.2.21. Make a tree diagram to list all ways to choose three onsets from eight possible ones where no two onsets are adjacent (next to one another).

Exercise 7.2.22. Suppose we are interested in all grooves of a bar of k onsets among n possible ones, if no two onsets can be adjacent. Show that there are no such patterns if $k > n/2$ and n is even or if $k > (n+1)/2$ and n is odd.

Exercise 7.2.23. For grooves in Exercise 7.2.22, show that there is exactly one pattern if $k = 0$, one such pattern if $k = (n+1)/2$ and n is odd, and two such patterns if $k = n/2$ and n is even.

Exercise 7.2.24. Let $c_{n,k}$ be the number of grooves described in Exercise 7.2.22. Show that $c_{n,k} = c_{n-2,k-1} + c_{n-1,k}$.

Exercise 7.2.25. Use Exercise 7.2.24 to check your answer to Exercise 7.2.21.

Exercise 7.2.26. Create a table, for the $c_{n,k}$ (described in Exercise 7.2.24) for n and k at most eight.

Exercise 7.2.27. Suppose now that we want to count grooves of a bar of k onsets among n possible ones, if no two onsets can be adjacent, and moreover we can't have the first and last onsets adjacent when you repeat the groove (that is, you can't choose the first and last possible onsets, either). List all ways to choose three onsets from eight possible ones subject to this rule.

Exercise 7.2.28. Let $d_{n,k}$ be the number of grooves described in Exercise 7.2.27. Find the values of $d_{n,k}$ when k is 0 or 1, when $k > n/2$, when $k = n/2$ (and n is even) and for n at most 3.

Exercise 7.2.29. Let $d_{n,k}$ and $d_{n,k}$ be as described in Exercises 7.2.24 and 7.2.27. Show that for $n \geq 4$ we have $d_{n,k} = c_{n-3,k-1} + c_{n-1,k}$.

Exercise 7.2.30. Count grooves of a bar of any number of onsets among eight possible ones if no two onsets can be adjacent, and moreover we can't have the first and last onsets adjacent when you repeat the groove (that is, you can't choose the first and last possible onsets, either).

Exercise 7.2.31. Suppose you flip a coin for each of 16 possible onsets—if the coin comes up heads, you play on the onset, and if it comes up tails you don't. (a) How many possible patterns are there? (b) What is the most likely number of onsets you would play? (c) About how many of the patterns would have between 6 and 10 onsets? Between 4 and 12 onsets? Between 2 and 14 onsets?

(Hint: Think of the process as selecting a random sample of 16 from a population whose proportion of "played onsets" is 0.5.)

Exercise 7.2.32. Suppose you flip two coins for each of 48 possible onsets—if both coins come up heads, you play on the onset, and otherwise you don't. (a) How many possible patterns are there? (b) What is the most likely number of onsets you would play? (c) About how many of the patterns would have between 9 and 15 onsets? Between 6 and 18 onsets? Between 3 and 21 onsets?

(Hint: Think of the process as selecting a random sample of 48 from a population whose proportion of "played onsets" is 0.25.)

Exercise 7.2.33. Listen to The Beatles' songs "A Hard Day's Night," "Here Comes the Sun" and "Something". Locate in each song instances of the patterns of threes against background patterns of fours.

Exercise 7.2.34. What is so deceptive about the opening guitar chords of The Beatles' "She's a Woman"? Explain what mathematical transformation(s) John Lennon, the rhythm guitarist, did to create the tension.

Exercise 7.2.35. Listen to the opening guitar riff of "Walk This Way" by Aerosmith. Can you decide what transformation the lead guitarist did to the first four notes to create the second four?

Exercise 7.2.36. Listen to Foreigner's "Head Games," where the title words are sung on beats 4 and 1 of two consecutive bars in the chorus. Listen to the chorus throughout. Do you notice any change? Can you explain what the composers did mathematically to play a head game with the listener?

Exercise 7.2.37. Find instances in the guitar parts of ACDC's "Back in Black" of groupings of threes against background patterns of fours.

Exercise 7.2.38. Grab a partner and ask him or her to clap regularly, counting 1, 2, 3, 4 repeatedly, and emphasizing beat "1" every time. Then you try to clap regularly, on the same beats, but count 1, 2, 3 and emphasize your beat "1." Listen to the effect. Try out other length patterns against one another, and predict for each pair of beating lengths the first point at which the emphasized beats will coincide, based on the least common multiple.

Exercise 7.2.39. (For musicians) Listen to the middle section (the *bridge*) of "I Want to Hold Your Hand." The chords are so very interesting yet fit the song perfectly. The chords to the bridge are Dm7, G, C, Am, Dm7, G, C, each a bar long, except for the C, which lasts almost two bars. The chorus has chords C, D, G, Em, C, D, G, each for half a bar, except for the last G, which lasts for a bar. Almost all popular songs (and parts of songs) have a *key*, which is a particular major or minor scale on which it is based. The Beatles changed keys in "I Want to Hold Your Hand," from the key of G for the verse and chorus, modulating to the key of C in the bridge. Can you figure out how transformations played a role in the chord structures? (Hint: The chord Dm7 consists of the notes DFAC. To read more, see www.jasonibrown.com/articles/iwthyh.pdf.)

Exercise 7.2.40. (For musicians) Create a riff of eight notes, starting with C4. Write them out as a list of numbers, in terms of the semitones up from C4. Then apply translations of the form $f(x) = x + a$ for some small numbers a. For example, if your riff is

$$C4, C4, Eb4, E4, G4, G4, A4, G4$$

you rewrite it as the number sequence

$$0, 0, 3, 4, 7, 7, 9, 7.$$

Then applying the translations $f_1(x) = x$, f_1, $f_2 = x + 5$, $f_3(x) = x + 7$, we get the sequences

$$0,0,3,4,7,7,9,7,\ 0,0,3,4,7,7,9,7,\ 5,5,8,9,12,12,14,12,\ 7,7,10,11,14,14,16,14$$

or, in terms of notes,

C4,C4,Eb4,E4,G4,G4,A4,G4, C4,C4,Eb4,E4,G4,G4,A4,G4, F4,F4,Ab4,A4,C5,C5,D5,C5, G4,G4,Bb4,B4,D5,D5,E5,D5.

Play the original sequence once (as a bar of eighth notes) and then the pattern of four bars over and over again. Try out some other kinds of functions (compression/stretches, reflections), to both the notes and their rhythms.

7.3   *Solving Musical Mysteries with MSI (Math Scene Investigations)*

While mathematics plays an intrinsic role in both the listening to and composing of music, it can also be used to uncover some musical mysteries, and that is precisely what I have done over the years. Sometimes mathematics can explain why we like the sounds and music that we do and sometimes it can unravel how music was created, when musicians want the mystery to continue or memories fade. The examples here are a smattering of the research I have carried out on applying mathematics to mysteries in popular music.

Why Are the Blues So Damn Good?

The *blues* are rather unique—some musicians spend their entire lives playing music in this genre, and others spend their entire lives listening to it. And the surprising thing is how similar many of the songs are. Most blues are 12 bars long (in either $\frac{4}{4}$ or $\frac{12}{8}$ time, so with four basic beats in a bar, each divided into two or three sub-beats), with a certain sequence of basic chords in a certain structure: once a key is chosen, the chord progression is most often

Figure 7.20: The great blues artist, B.B. King, and his guitar, *Lucille*.

- the I chord for four bars,

- the IV chord for two bars, the I chord again for two bars, then

- the V, IV, I and V chords, each for one bar

(the choice of whether the chords are major or minor can vary from blues to blues). And the sound of the voice, both in note choice and timbre, is quite moving. So the question remains— why are the blues so good? That is something we'll try to answer mathematically in this section.

We've seen that all complex tones are made up of pure tones, represented by sine waves, by addition. But what we hear is the sum of many (possibly thousands of) such sine waves. Like the ingredients in a cake recipe, they all add together and in the end it seems impossible to tell what and how much of each went into the final product. Amazingly, there is a mathematical process that uncovers the "ingredients" in our musical recipe, and it is called a **Fourier transform**, which decomposes a musical sound into its component frequencies (and their amplitudes). The process is due to the French mathematician and scientist Joseph Fourier (1768–1830), who was investigating heat transfer

Figure 7.21: Mathematician Joseph Fourier.

in physics. Fourier transforms have many applications in science, but they are crucial to understanding music.

The essentials of Fourier transforms require a background in calculus, so we won't discuss any details. But the key thing is that it is a process that can be programmed (and has been) that can produce for a complex tone its **spectrum**, that is, its component frequencies with their amplitudes. While the process is involved, it is fascinating to note that each of us has some anatomical ability to carry out Fourier transforms subconsciously—with practice, we can listen to a sound played by an orchestra and hear the different notes that are being played!

Blues singers sound so compelling that I wanted to know what went into their singing. What notes do they sing? And what are they doing when they growl out a note? It seems perhaps that answers are unlikely—what they do is just *magic*. But mathematics can help us look behind the curtain.

It seems pretty clear that singers have to sing notes from the equally tempered scale (or as close to that as possible) when they are singing with instruments that are equally tempered, as otherwise the singer would sound out of tune. But many blues singers sing "a capella," that is, without instruments, and then the field is more wide open. Do they choose to sing the notes from a equally tempered scale anyway, just from practice? Or will they naturally gravitate to notes from other scales?

Being The Beatles aficionado that I am, I tried examining The Beatles' catalogue, looking for a long held "bluesy" note without any influence of background instruments. I found one such example, a cover of a song called "Mr. Moonlight." John Lennon screeches a long "Mis-TER" at the opening. The note of "TER" is an A♯ while the song is in the key of F♯.

I took a portion of the opening A♯ that John sings—Figure 7.23 shows its spectrum, with the frequencies along the horizontal axis and the amplitudes along the vertical axis. Vertical bars to the right correspond to higher pitches, and the higher a bar, the louder its frequency is. Looking at the spectrum (which I created from data points), we see that the A♯4 note that John sings is a

One free program, both for Windows and Macs, is *WavePad*, a sound editor. Once you select a sound, you can choose to do a Frequency Analysis, or FFT (*Fast Fourier Transform*), where you can move your cursor over a peak to see both the corresponding frequency and nearest musical note to that component of the spectrum.

The main scale for the blues (and basic rock 'n' roll) is the pentatonic scale, with semitone sizes 3, 2, 2, 3, 2. So, for example, in the key of A, the blues scale is A,C,D,E,G,A. The *flat third* would be the "C" note, the minor third up from the key note (or *tonic*), A, and is most tension-filled of the song, especially when the chords being played are major chords. Of course, other notes from the major or minor scales, such as the major third and sixth, are often added in to "spice" things up.

Figure 7.22: John Lennon.

little flat for the key. The F♯ below the A3, F♯4, has frequency 370 Hz, so we'd expect the A♯4, four semitones above that, should have a frequency of about $370 \times 2^{3/12} = 466$ Hz; in contrast, an A4 would have a frequency of 440 Hz. The spectrum shows that John sings a 460 Hz note. Is Lennon just singing a bit out of tune?

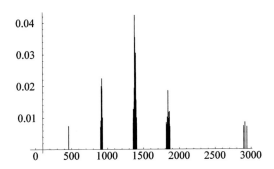

Figure 7.23: Opening note of *Mr. Moonlight.*

The answer I think is more interesting. Undoubtedly, just before recording, John heard/played the tonic chord, that is, the F♯. Good blues (and rock) singers (including John Lennon), when performing without instruments, sing not the equally tempered major third, but the just third (see Exercise 7.1.31), which has a ratio of 5/4 (one of those nice small fractions). The just third above F♯4 has frequency $\frac{5}{4} \times 370 = 462.5$ Hz, which is just about what John wails.

The spectrum can help answer also how great blues and rock sings "growl" out a note in that gruff, powerful voice. To investigate, I need to compare such a sound to its opposite, a sweet melodic note. So to contrast with John howling his opening note of "Mr. Moonlight," let's look at the spectrum of Paul McCartney's opening sweet note on "Hey Jude."

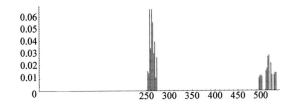

Figure 7.24: Opening note of *Hey Jude.*

The spectrum in Figure 7.24 is made up of a number of pure tones, but essentially there is one loud frequency at about 261 Hz—the fundamental, very close to middle C—and a second, around 522 Hz, which is twice the first frequency (that is, the first harmonic). There are no other harmonics that appear at any significant amplitude.

The spectrum for John's harsh, bluesy opening of "Mr. Moonlight . . . " is quite different. There you can see a high peak at around 920 Hz (which is almost an A$\sharp$—see earlier), with a second even stronger frequency at approximately 1380 Hz. The third frequency produced by John, at 1380 Hz, is 3/2 times the peak at 920 Hz, so in this gruff sounding voice, John is manipulating his throat to emphasize the two notes simultaneously (in addition to the first harmonic at an octave above the original) that form a perfect fifth—rock music's *power chord*—with the top of the fifth being louder than the lower frequency—that gives his voice its strength.

Figure 7.25: Paul McCartney.

The chord progressions of so many rock songs are based on the blues: Chuck Berry's *Johnny B. Goode*, Elvis Presley's *Hound Dog*, Little Richard's *Lucille* and *Long Tall Sally*, The Beatles' *The Word*, Led Zeppelin's *Rock n Roll*, Loggins and Messina's *Your Mama Don't Dance*, and classics like *Rock Around the Clock*, *Dizzie Miss Lizzie*, *Bad Boy* and *Kansas City*. Songs like *Day Tripper* are very interesting in the way they break out of the expected blues progression that they so carefully set up.

Like its masters, the blues just get better with age. I have yet to see any reasoning why the blues are the way they are—why they have the chords they have, and why the chords last as long as they do. But the feeling the blues chord progression generates is so powerful, so eternal, so right, that I sense there must be some underlying mathematics. And what I've discovered is what I think is the mathematics behind the beauty of the blues.

As mentioned earlier, the basic blues chord sequence (or *progression*) is the following: the *tonic* chord (that is, chord I in the key) lasts for four bars (each consisting of four beats), followed by chord IV (the *subdominant*, on the fourth of the key) for two bars, followed by chord I for two bars, and then chords V (the *dominant*, on the fifth of the key), IV, I and V, each for one bar (in the following *chord chart*, a slash / denotes a strum of the chord whose name was last shown):

```
I   /  /  /  |  /  /  /  /  |  /  /  /  /  |  /  /  /  /  |
IV  /  /  /  |  /  /  /  /  |  I  /  /  /  |  /  /  /  /  |
V   /  /  /  |  IV /  /  /  |  I  /  /  /  |  V  /  /  /  |
```

Sometimes the chords used are the minor rather than the major variant, and you will find examples of the blues with other small modifications, but it is surprising how often the blues sticks to the basic pattern.

In its essence, the blues are *thrilling*. So to try to figure out why the blues are so exciting, I tried to model the blues mathematically, imagining what analogy I might draw to real life. I thought back to my childhood. Remember how exciting it was to go on a roller coaster? Up and down, peak to valley to peak again. The tracks were laid out to maximize the tension and release. The roller coaster is the perfect paradigm of a ride meant to excite. Then I thought, how can I create a chord progression, that is, a sequence of chords, that accurately models the thrilling ride of a roller coaster?

> When you try to model a situation mathematically, it often helps to think of analogies to other objects that can be modeled mathematically.

In my ideal mathematical model of a roller coaster, I want to have three equal sections for the ride—the beginning, middle and end. I also want the excitement to build from one section to the next. The easiest way to accelerate is to double the sense of movement from one section to the next. So if the chord progression stays on one chord for the beginning section, it should have two chords equally spaced for the middle and four equally spaced chords for the finish. Working backwards, if we take the smallest segment for a chord change to be one bar, then we should make the finish four bars, each with a different chord, the middle with two chords, each lasting two bars, and the beginning part with just one chord lasting for four bars. What do we have in total? Twelve bars for our musical roller coaster! See Figure 7.27 for a view of the musical roller coaster from the top.

A big part of mathematics is *visualizing*, and it is an art to

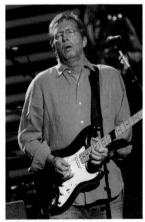

Figure 7.26: Eric Clapton, one of the great blues guitarists.

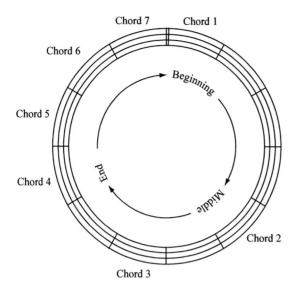

Chord 7 Chord 1
Chord 6
Chord 5
Chord 4
Chord 3
Chord 2

Beginning
Middle
End

Figure 7.27: Blues roller coaster.

figure out how best to represent information. I'm going to mathematically model my blues chord roller coaster with a graph! I'll replace each chord by a vertex and join two by a edge when they are played one after the other (I label the points/chords as $v_1, v_2, \ldots, v_7$). This is exactly what you want to do mathematically—keep what is essential and hide the rest.

The graph in Figure 7.28 is C_7, the **cycle** on 7 points. The value of the mathematical model rests on whether we can model the assignment of chords to the points. We can think of each chord as being a different "color" that we can assign to the points (namely, the chord that will be played at that time in the blues). The proviso is that adjacent vertices must receive different colors (as otherwise, we won't notice the change). Such assignments of colors to points in a graph has been well studied over many years—they are called *colorings* (see Exercise 3.2.26).

Now we can see that there can be no blues roller coaster with only two chords, as we can't properly color C_7 with only two colors. For if the colors we tried to use are red and blue, then if the first point v_1 say is colored red, then v_2 must be colored blue, v_3 colored red, v_4 colored blue, and so on, and we find that v_7 gets colored red, a problem, as v_1 is colored red already and

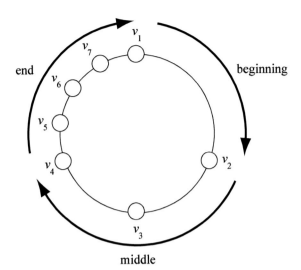

Figure 7.28: Blues roller coaster as a graph.

joined to it. We get a similar problem if we start by coloring v_1, so we can't color with only two colors—there is no blues roller coaster with only two chords.

Therefore, we see immediately that we will need at least three chords in our roller coaster chord progression, and indeed three chords (our "colors") will suffice. We'll build our most basic musical roller coaster by sticking to our three strongest chords: the I, IV and V chords, so we take these as our colors for our chord progression. Musically, the I chord provides the least tension (it is the root chord of the key), while the V chord provides the most—not all colors are created equal here. Now let's talk more about the shape of the roller coaster. We'll start off at base level; musically, this would be the tonic chord, that is, the major chord that defines the key. Thus vertex v_1 gets color I. Vertex v_2 is our first climb (in the middle section of the roller coaster), and we should save the most exciting color (V) for the end of the ride. That means v_2 and v_3 must be colored IV and I, respectively.

Now for the big finish. We save out biggest climb for the beginning of the finish, where we jump up to chord V for chord 4, that is, we use our most exciting color, V, for v_4. A good roller coaster would prolong our descent back down to earth, so chord 5 should not be the tonic I, which leaves v_5 to be colored with

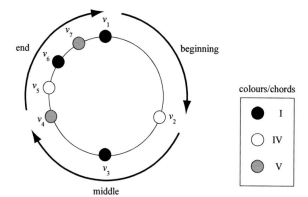

Figure 7.29: Finished blues roller coaster as a graph.

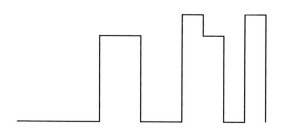

Figure 7.30: An abstract one-time trip around the blues roller coaster.

IV. Dropping down to the ground, we see that v_6 should be colored with the tonic chord, I. We save one great thrill for the end, by climbing back up to the top, with chord 7, so that the final vertex, v_7, of our seven cycle chord progression, is colored with V. Our unique roller coaster coloring (see Figure 7.29) is indeed the standard 12 bar blues, and having completed the ride, another can begin!

Unraveling a Musical Mystery with a Hard Day's Math

If you asked 100 rock guitarists to name the most famous guitar chord in rock 'n' roll history, I'm pretty sure that at least 99 would name the opening chord to The Beatles' "A Hard Day's Night." In fact, if you play that glorious opening "twang" for anyone over 40, it would be practically impossible not to sing "It's been a . . ." Movie reviews of "A Hard Day's Night" always make mention of it; Roger Ebert[1] (see www.suntimes.com/ebert/ebert_reviews/1999/10/ night1029.html) wrote "Perhaps this was the movie that sounded the first note of the new decade—the opening chord on George Harrison's new 12 string guitar."

[1] R. Ebert. *A Hard Day's Night* (movie review). 1999

The Beatles and their producer, George Martin, knew exactly how important the opening chord was, to both the album and their first film (in *black and white*!). In fact, Martin was quoted[2] as saying, "We knew it would open both the film and the soundtrack LP, so we wanted a particularly strong and effective beginning. The strident guitar chord was the perfect match."

[2] M. Lewisohn. *The Complete Beatles Recording Sessions.* Doubleday, Toronto, 1988

And yet as important as this chord is to rock 'n' roll, there has been much controversy over the years as to how it was played. Because of our innate ability to hear frequencies in chords (via our brain's built-in Fourier transform capabilities and a basic knowledge of harmony), many guitarists have come up with their own ideas of how the chord was played. Three different well-known versions are

- G,C,F,Bb,D,G (an easy to play barre chord, with a finger across all guitar strings at the third fret, see Figure 7.31),

- G,D,F,C,D,G (another common version, and perhaps a bit closer to the real thing, see Figure 7.32), and

Figure 7.31: Version 1 of The Chord. The TAB notation for guitars is fairly mathematical as well—the six line staff has a line for each guitar string and the number on a string indicates where to place a finger.

Figure 7.32: Version 2 of The Chord.

- the chord shown in Figure 7.33, which has three of The Beatles playing on it.

Figure 7.33: Version 3 of The Chord.

(The last one is from "The Beatles Complete"[3]).

But how should someone choose between them? Just by his or her own ears? As good as some people are at detecting notes, everyone sometimes misses frequencies that are there and inserts frequencies that are missing. So what is one to do?

[3] H. Kubo Y. Hagino and G. Sato. *The Beatles Complete Scores*. Hal Leonard, Milwaukee, 1993

This is the point where mathematics can come to the rescue—Fourier transforms provide a way to find out what frequencies are present and at what amplitudes. I took a small sample of the chord and ran it through a Fourier transform procedure on my computer. Tens of thousands of frequencies appeared in the list, but fortunately, the Fourier transform also gives out the amplitudes, and I know that the notes played are among the loudest sounds, and hence among those with the largest amplitudes. Considering only the frequencies with the largest 48 amplitudes (a natural breaking point, from the data), I got the beautiful table in Figure 7.34, whose frequencies range about from 110 Hz (cycles per second) to 3158 Hz. Not all of the frequencies are likely to be notes that George played, but some would be. The plan is to figure out which ones are.

Freq. (Hz)	Ampl.	Freq. (Hz)	Ampl.
110.34	0.0600967	299.494	0.0298296
145.619	0.025485	392.57	0.0309716
148.621	0.0264278	438.358	0.0286329
149.372	0.0656018	524.678	0.0680974
150.123	0.175149	587.73	0.020613
174.142	0.0275547	588.48	0.0310337
174.893	0.0380282	589.231	0.0231753
175.643	0.0407103	785.141	0.0323532
195.159	0.0405164	786.642	0.0251928
218.428	0.0448308	787.393	0.0268553
261.964	0.0302402	960.784	0.0228509
262.714	0.0234502	981.801	0.02242

Figure 7.34: The 48 frequencies of largest amplitude.

Working with frequencies is OK for mathematicians and scientists, but musicians play notes. So I needed to convert from frequencies to musical notes. Using some more high school mathematics (see Exercises 7.3.7 to 7.3.9), I converted the 48 frequencies into semitones away from note A3 (with a frequency of 220 Hz), and I got:

−11.9466,	−7.1437,	−6.7904,	−6.7031,	−6.6164,	−4.0469,	−3.9724,	−3.8983,
−2.0742,	−0.1241,	3.0224,	3.0719,	5.3403,	10.0254,	11.9353,	15.0472,
17.0118,	17.0339,	17.0560,	22.0254,	22.0584,	22.0750,	25.5205,	25.8951,
27.0719,	29.1659,	30.5752,	30.9449,	31.0238,	31.0337,	33.0990,	34.6990,
35.9056,	41.0780,	41.1330,	41.1385,	41.1439,	41.1604,	41.1659,	41.1714,
43.0042,	43.0091,	43.7514,	43.8127,	45.7080,	46.0626,	46.0667,	46.1244

Now something should hit you—aren't notes a whole number of semitones away from others? Indeed! But some of the numbers we get are not integers, so what gives? The answer is that The Beatles' producer ought to have told the lads to retune their guitars, as they were slightly out of tune (the final take was number 9, after a number of runs through the song, and guitars naturally lose their tuning over time). One of the fascinating mysteries that we have uncovered in the chord already is that The Beatles were slightly out of tune!

Of course, they weren't *so* out of tune, so what we can do to handle the tuning is to round each number of semitones to the nearest integer—that takes them to the nearest note:

A_2,	D_3,	D_3,	D_3,	D_3,	F_3,	F_3,	F_3,
G_3,	A_3,	C_4,	C_4,	D_4,	G_4,	A_4,	C_5,
D_5,	D_5,	D_5,	G_5,	G_5,	G_5,	B_5,	B_5,
C_6,	D_6,	E_6,	E_6,	E_6,	E_6,	$F\sharp_6$,	$G\sharp_6$,
A_6,	D_7,	D_7,	D_7,	D_7,	D_7,	D_7,	
E_7,	E_7,	F_7,	F_7,	G_7,	G_7,	G_7,	G_7

A lot of the notes are not surprising—many of them appear in the three versions of the opening chord that we have seen. The mathematics confirms what people's ears tell them! But again another surprise arises—all of the three versions have a low G2 in them (the third fret on the lowest string of the guitar), but the note is absent from the data. We can immediately conclude that *all* of the three versions are wrong!

It is well known that on this song George Harrison played a brand new Rickenbacker 12-string guitar, whose strings are, from lowest to highest, E2 E3 A2 A3 D3 D4 G3 G4 B4 B4 E5 E5. Realizing that strings strummed at the same time on a guitar would have roughly the same loudness, we reconsider the frequencies and their amplitudes in Figure 7.34. There we find that one of the D3's has much higher amplitude (i.e., is much louder), with a value of 0.175. It seems clear that this must come from an instrument that plays only one note at a time—Paul McCartney's Hofner bass. We can start pairing off octaves such as the A2 and A3 as supposedly coming from Harrison's twelve-string guitar.

But there is a BIG problem. With a D3 accounted for on McCartney's bass, there are three other D3's to account for. And the two F3's pose even a bigger dilemma. No matter how Harrison played an F3 on his 12-string guitar, he would have to have played an F4 at the same time, and there is no F4 there. Is the math in trouble?

I lost some sleep about this, but all of sudden a thought hit me—what if the assumption that it was only The Beatles playing on the opening chord was *wrong*? I knew that The Beatles' record producer George Martin doubled on piano Harrison's solo in the song. Perhaps he had added in a piano chord to the opening? Recall that pianos have hammers under each note that strike one, two or three identically tuned strings, depending on where the key is (bottom, middle or top of the piano). The three F3's could all have arisen from an F3 piano note! The frequencies of the three F3's were indeed slightly different, which added credence to my piano hypothesis.

Now back to the problem of the three D3's. The bottom ten pitches on a piano are just single strings which change over to pairs of strings, and somewhere near C3 triples of strings appear under each note. A quick trip to a local piano store and counting strings under keys assured me that indeed some grand pianos (those of medium length) change over from two to three strings right around D3! I concluded that two of the D3's were played on the piano (as well as knowing that the grand piano in Abbey Road studios was of medium length!) Further deductions led to the discovery that George Harrison played

A2 A3 D3 D4 G3 G4 C4 C4

on his 12-string guitar, markedly different from any of the other versions roaming around the web. With some more thought, I came up with the chord shown in Figure 7.35. My version of "the chord" was published in *Guitar Player Magazine* and traveled virally a number of times around the internet. Mathematics and music—that's a sweet sounding combination!

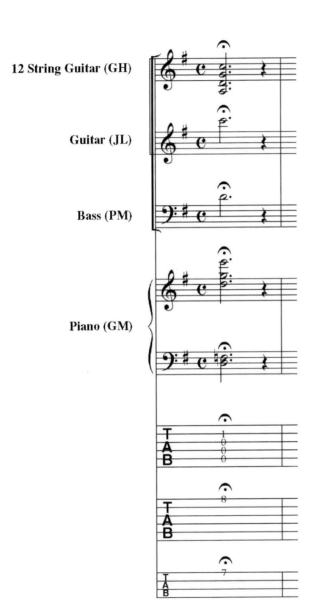

Figure 7.35: The chord!

Exercises

Exercise 7.3.1. Explain why any odd cycle graph (that is, a cycle with an odd number of points) can never be colored with two colors.

Exercise 7.3.2. Can an even cycle graph (that is, a cycle with an even number of points) be colored with two colors? If so, in how many different ways?

Exercise 7.3.3. A **path graph** P_n on n points can be formed from the cycle graph C_n on n points by removing one edge. Draw a picture of P_4 and P_5. If you had the points represent chords that are to be played, as was done for the blues, explain why a path graph may the right model if we only play the blues once.

Exercise 7.3.4. Explain why P_n can always be colored with two colors. In how many ways can you do so?

Exercise 7.3.5. (Harder) In how many ways can the path graph P_n be colored with three colors? With k colors?

Exercise 7.3.6. (Harder) If you don't have any restrictions on the assignment of colors to the points of C_7, how many ways can you color the points with three colors?

Exercise 7.3.7. Let's fix a certain note, say A3, with a frequency of 220 Hz. If a note is i semitones away from A3 (i is positive if the note is higher, negative if the note is lower), explain why its frequency is given by

$$f = 220 \times 2^{i/12}.$$

Exercise 7.3.8. (Continuing with Exercise 7.3.7) Suppose we reverse things and are given the frequency f and want to find the number of semitones i we are away from A3. We can use **logarithms** to help us. Choosing any positive number a you like as a base, $\log_a x$ is the power you raise a to get x. There are some very useful rules for logarithms, with respect to multiplication, division and powers, namely, for positive number a, x and y and number c:

- $\log_a 1 = 0$,
- $\log_a a = 1$,
- $\log_a(xy) = \log_a x + \log_a y$,
- $\log_a(x/y) = \log_a x - \log_a y$, and
- $\log_a x^c = c \log_a x$.

Use these rules to show that if $f = 220 \times 2^{i/12}$, then $i = 12 \log_2(f/220)$.

Exercise 7.3.9. Use the previous exercise to check the first and last numbers in the table on page 385 that give the number of semitones the corresponding frequencies in Table 7.34 are away from A3.

Exercise 7.3.10. Logarithms convert multiplication into addition. Explain what this implies about stacking two intervals.

Exercise 7.3.11. Logarithms convert powers into multiplication. Explain what this implies about repeating intervals.

Exercise 7.3.12. Old recording machines used magnetic tape for recording. It's well known (and not hard to see!) that the frequency of a recorded note is proportional to the tape speed—if you double the tape speed, you also double the frequency of the recorded tone. Some early tape machines had two speeds—7.5 inches per second and 15 inches per second, with the quality better at the higher speed. If you recorded a frequency of 335 Hz at 7.5 inches per second and played it back at 15 inches per second, what frequency would be heard?

Exercise 7.3.13. If you recorded a frequency of 335 Hz at 15 inches per second and played it back at 7.5 inches per second, what frequency would be heard?

Exercise 7.3.14. What advantage would there be to learning a musical passage by recording it at 15 inches per second and playing it back at 7.5 inches per second?

Exercise 7.3.15. In the mid-1960s *varispeed* tape machines were created that allowed one to set the tape speed directly, without having only 7.5 and 15 inches per second as your only choices. If the tape speed is 15 inches per second and you want to move the song up by 4 semitones, what should the new tape speed be?

Exercise 7.3.16. If the tape speed is 15 inches per second and you want to move the song down by 5 semitones, what should the new tape speed be?

Exercise 7.3.17. The Beatles recorded "Strawberry Fields Forever" twice, once in the key of A at a slow speed, and then faster, in the key of B. John liked the first half of the slow version and the second half of the second. What did the producer, George Martin, have to multiply the tape speed of the slower first take by in order to bring it up to the key of B as well?

Exercise 7.3.18. The two takes of "Strawberry Fields Forever" described in the previous exercise had different *tempos* as well—the first was played at 85 beats per minute while the second was played at 107 beats per minute. The rhythm (as a number of beats per minute) is also proportional to the tape speed. What would George Martin have to multiply the tape speed of the slower first take by in order to bring them up to the same tempo? (As people are more sensitive to changes in frequency than changes in rhythmic tempo, he would reset the tape speed of the first to match the key of the second rather than the tempo of the second).

Exercise 7.3.19. Listen to George Harrison's famous solo in "A Hard Day's Night." The very fast sixteenth notes—what do you observe about the patterns?

Exercise 7.3.20. Try building your own musical blues roller coaster, perhaps by having four equal sections. Try out the results musically, if you can!

Exercise 7.3.21. Using a software program like WavePad, examine the spectrum of musical files at `www.jasonibrown.com/mla`, and see, for each, what notes you think make up the chords.

And Now the Rest of the Story . . .

George Martin did return again, without the lads, but with a sound engineer. He must have sat for quite a while at the piano, trying out different chord clusters to add in to the opening chord. The chord that George Harrison played on his guitar is very unusual, but easy to play, with lots of open strings. McCartney's note shows that The Beatles thought of the chord as a dominant V chord, a natural (tension-filled) one to drive into the song. But the opening needed to be MASSIVE. When he was in his teens, Martin thought he might be a classical composer like Serge Rachmaninoff, and his mind began to fill with the huge chord clusters favored by Rachmaninoff. Finally, he struck the right one—he knew it once he heard it. Listening to the combined Beatles– Martin chord was a pleasure all its own. Martin even chuckled at adding in an "F" that was missing in The Beatles' version, the very note that figured prominently in the song and its jangling ending. Perfect!

Later, the boys had to agree. Even if it added in a classical rather than rock dimension to the opening chord, it didn't matter—all that was important was that the chord was <u>it</u>. And George Harrison, who had trouble with the brilliant solo he heard in his head but couldn't quite play—it had Berry-esque groupings of threes against fours, but with sixteenths instead of the usual eights. George was determined to get it down, returning to the studio later with George Martin to play the solo in unison onto a single track of tape. Precise and determined as always, Harrison insisted that the tape speed be cut in half and both he and Martin played the solo he had worked out, down an octave, just to get it right.

The Beatles may have been rock 'n' rollers, but they were musicians first. The hard part would be figuring out how to revoice the chord for playing the song publicly, as there would be no piano (and no George Martin) on stage with them. But the silence that filled the room after hearing the opening chord played back told everyone that this sound would be indelibly transfixed to the song, and indeed their careers.

4 *Review Exercises*

Exercise 7.4.1. For each of the following, convert the frequency in kilohertz to hertz.
  (a) 2.5 kHz   (b) 0.50 kHz   (c) 25 kHz   (d) 96 kHz
   Which are within the range of human hearing?

Exercise 7.4.2. Write out the formula for a corresponding sine curve with frequency 660 Hz, amplitude 1.25.

Exercise 7.4.3. Find the frequency of F below middle C (it may be helpful to compare it to A 220 Hz).

Exercise 7.4.4. The melody of the song "Three Blind Mice" is

$$E4,D4,C4,E4,D4,C4,G4,F4,E4,G4,F4,E4,G4,C5,C5,B4,A4,B4,$$
$$C5,G4,G4,G4,C5,C5,B4,A4,B4,C5,G4,G4,F4,E4,D4,C4,E4,D4,C4.$$

If we denote middle C by 0, write out the sequence in terms of the number of semitones each note is up from middle C.

Exercise 7.4.5. Transpose the melody of "Three Blind Mice" up a perfect fourth (that is, five semitones) by first adding five to each number in the corresponding sequence (see the previous problem) and then transferring the sequence back to notes.

Exercise 7.4.6. How many eighth notes are there in five bars of three quarter notes each?

Exercise 7.4.7. Let's fix a certain note, say C4, with a frequency of 262 Hz. If a note is i semitones away from C4 (i is positive if the note is higher, negative if the note is lower), explain why its frequency is given by

$$f = 262 \times 2^{i/12}.$$

Exercise 7.4.8. If you recorded a frequency of 450 Hz at 15 inches per second and played it back at 7.5 inches per second, what frequency would be heard?

Exercise 7.4.9. If the tape speed is 15 inches per second and you want to move the song up by five semitones, what should the new tape speed be?

Exercise 7.4.10. It's known that George Martin, The Beatles' producer, doubled on piano George Harrison's guitar solo on the same recording track (that is, George Martin played the same notes on piano as George Harrison played on guitar, at the same time). When I examined the frequencies of a G note from the solo, I found three G3's and one G4. What does that tell you about the recording of the solo?

Late Night Mathematics—Humor and Philosophy

Kurt Gödel was born in 1906 in Brünn, Austria (which is now Brno in the Czech Republic). A flair for the arts ran in his family—his grandfather, Joseph Gödel, was a well-known singer. But Kurt was inclined in other ways. He was known in his family for his inquisitive nature, always wanting to know the reasons for things. In school he excelled, especially in mathematics, and attended the University of Vienna at age 18, taking courses in mathematics, physics and philosophy.

*Gödel was interested in number theory, but a seminar in logic captured his attention. The famous philosopher Bertrand Russell had written a book, "Introduction to Mathematical Philosophy," whose goal was to find a list of rules or "axioms" from which all mathematical truths could be derived. The plan was accentuated by a lecture by world renowned mathematician David Hilbert, who raised the problem of whether certain logical systems (or **theories**) were **consistent** (that is, had no inherent contradictions!), and whether they were **complete**, that every sentence was either provably true or provably false.*

Gödel proved in his 1929 Ph.D. thesis (at age 23) that a certain logic (called first-order logic) was indeed complete. But he still pondered the gauntlet that Russell had thrown down—is there a set of axioms for mathematics from which every mathematical truth could be proved?

We've come to the end of the book. We've seen a lot ways in which mathematics plays an important role in our lives. It seems that mathematics always has the right answers, and it exists, independent of us. But in fact, there are many philosophical questions that surround mathematics. And that is the "note" we want to end off on.

Laughing with Mathematics

MATHEMATICS MAY SEEM SERIOUS, BUT IT CAN BE PLAYFUL AS WELL. Mathematicians are always open to puzzles, ones that are difficult, impossibly hard or even seemingly contradictory. The more thought involved, the better the payoff.

The Lighter Side in Apparent and Real Contradictions

Let's start with some errors masquerading as paradoxes. First, let's prove that $0 = 1$! First suppose that $x = 0$. Then $x^2 = 0$ so from this and $x = 0$ we find that $x^2 = x$. Clearly $x = 1x$, so from

$$x^2 = 1x$$

we cancel out an x on both sides to get

$$x = 1$$

and so, as x started out as 0, we find that

$$0 = 1$$

as promised!

The issue is that we can only cancel out a number on both sides of an equation by division if what we are dividing is known not to be 0. But we started with $x = 0$, so we <u>can't</u> cancel out 0 on both sides of $x^2 = 1x$. Doing so leads to the apparent contradiction.

There are other similar ones, having to do with simple mathematical operations. If we start with say $x = -5$ and square both sides, we get $x^2 = 25$. But then taking square roots, we find that $x = \sqrt{25} = 5$, so $-5 = 5$, again a contradiction. The error here lies in the fact that squaring and taking square roots are not quite inverses of one another—when you take square roots, there are often two choices that are available, one positive, one negative, and omitting one from consideration leads you down a trail of no return.

OK, sometimes contradictions arise from mistakes that can be corrected. But some are more involved. For example, think about the following: Suppose you are a librarian in a large library. You decide to make a master book to keep track of all books in the library. Your rule is as follows. You look inside any given library book. If its title does not appear on any page within the book, you write down the title of the book in your master book; otherwise, you don't. So, for example, if a book is titled "The Humor of a Mathematician," and after searching through the book, you can't find the title, as a phrase, anywhere in the book, you write the title "The Humor of a Mathematician" on a page in the master book. Here is the dilemma—the master book, whose title is "Master Book," is to be added to the library catalog. Do you write its name in the master book? If you do, you shouldn't (as its name appears on a page in the book), but if you don't, you ought to!

Here the problem is not so easy to deal with. There are no apparent math errors or errors of reasoning. And yet we have a full blown paradox on our hands! What do we do? Exactly such an issue led mathematicians to think more deeply about the collection of objects they deal with and whether statements can talk about themselves.

In the same vein, here is another paradox. Some numbers can be described with a small phrase. For example, 2 can be described by the phrase "the smallest prime number" and 6 by "the first number that is the sum of all of its positive divisors except itself" (as 6 has divisors 1, 2, 3 and 6, and $6 = 1 + 2 + 3$). Think about all the numbers that can be "tweeted," that is, that can be expressed by a sequence of say 140 characters (including spaces). Our description of 6 contains 78 characters, so 6 would be one of the numbers under discussion. Clearly not all whole numbers can be described, as there are infinitely many of these, and there are only finitely many sequences of 140 characters (including spaces, there are $26 + 1 = 27$ possible characters, so at most 27^{140} many such sequences could possibly define numbers—a huge number, but finite, nonetheless). So if not all whole numbers can be described within 140 characters, there has to be a least one—let's call it m. But the phrase "the least number that cannot be described in one hundred and forty characters" has just 76 characters, and describes m, the least number that can't be described within 140 characters. So we have a problem—we just described within 140 characters a number that can't be described in this way! Again, the problem lies with a sentence that refers to itself.

Here is just one more. Suppose you own a hotel, and you've gone the extra distance and built infinitely many hotel rooms, numbered, of course, 1, 2, 3, and so on, forever. Someone lands exhausted at midnight and asks for a room. You tell him, "Sorry, there aren't any rooms available. I have a guest in every room." Your clever straggler says to you, "I have an idea—why don't you ask everyone to move over a room? That is, ask the guest in room 1 to move to room 2, the guest in room 2 to move to room 3, and so on? Clearly every guest will still have a room, and I can crash in empty room 1." You think about it, and agree. So what gives? How can the hotel be full and still have room for another guest? The paradox here caused mathematicians to think about how one adds numbers to infinity.

> Numbers that are the sum of all of its positive divisors except themselves are called **perfect numbers**. The next perfect number after 6 is 28.

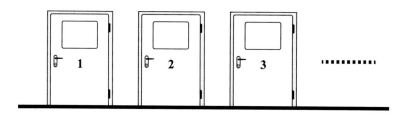

Figure 8.1: A hotel that can always accommodate one more.

Climbing the Levels

The problem with paradoxes like the Master Book problem is that items or sentences refer back to themselves, like the sentence "This sentence is false." If the sentence is true, then it is false, and if it is false, then it is true, so you get into a vicious circle of reasoning. One way out for logicians is to set up levels of reasoning, and objects are not allowed to talk about themselves— such sentences are only allowed at the next level up.

And the next level up for reasoning is called a **meta level**, where you can talk about what you can express in the previous level. For example, on a "basic" level (though it may not be basic at all!) you can carry out mathematical proofs, but at the meta-level, you can talk about what kind of sentences you can and can't prove.

Meta-level thinking can lead to humor. In songs, you are often expected to be empathizing with the singer, right in the song there with her or him, but some of the best songs require you to step up a level and look down on the singer. For example, in Dire Strait's *Money for Nothing*, the words on first listen seem offensive, but you have to realize that you are not to agree with the singer, but step out of the song and realize that the singer is a bit of an idiot! Randy Newman's early songs sometimes required the same type of listening.

There was an old variety TV show called the *Carol Burnett Show*, and I remember a skit with comedy great Carl Reiner, where he was a German general. He held up a round bomb in the skit, but looked at the audience and let them know that he thought the skit was bombing as well. The audience went wild—they weren't used to seeing an actor step up a level, out of the acting, and recognize not only the audience, but how the skit was going over. This meta-level thinking connected with the audience.

Figure 8.2: Comedy greats Mel Brooks (on the left) and Carl Reiner (on the right).

And his good friend, Mel Brooks, wrote a movie called *Blazing Saddles*, a comedic Western, where at the end, a brawl between the townsfolk and outlaws overflows in the movie into the movie studio's lot, with one of the characters leaving the movie lot to take in the very movie he's in, *Blazing Saddles*, in a theater! So many levels to think and laugh about!

Delighting in Ambiguity

Usually ambiguity, which occurs when something can have more than one meaning, is a bad thing in real life. If a significant other offers to play you a song he or she wrote and you respond, "I'd like nothing better," the ambiguity of your statement is likely to end up with you sleeping in the doghouse, at least for a few days. But in the arts, and especially in humor, ambiguity is a good thing—it allows your mind to play the phrase/joke/situation over and over again, switching back and forth between meanings, continuing the fun. Puns are exactly this type of joke—"I couldn't figure out why he launched the math book toward me, but suddenly it hit me."

In art, ambiguity is one way to keep viewers attention. When looking at the famous picture in Figure 8.3, you alternate between seeing two faces, or a vase in the center. The picture is ambiguous, and indeed you can see the two faces or the vase, but not both at the same time.

Figure 8.3: The urn and faces paradox, known as **Rubin's vase**, developed in the early 1900s by psychologist Edgar Rubin.

In mathematical logic, one can talk about different interpretations (or logical models) of sentences. For example, geometry, as described by the Greek mathematician Euclid, had the following rules or **axioms**:

- A line can be drawn through any two points.

- A line segment can be extended forever.

- For any given line segment, a circle can be drawn with one end of the segment as its center and the line segment as its radius.

- All right angles are congruent, that is, you can shift one over (by an **isometry**) to land up on the other).

He also put forward one more axiom, his **parallel postulate**. Together with his other axioms, it can be written in the following form, called **Playfair's axiom** (see Figure 8.4):

- Given a line L and a point p not on the line, there is exactly one line through p and parallel to L.

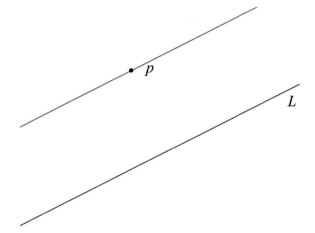

Figure 8.4: Playfair's axiom

The rules should seem all quite natural. What I want to point out is that the terms "point" and "line" are not defined, although what Euclid and other geometers were thinking of was the usual interpretation of points and lines. But when you try to prove things from axioms, what you prove must be true, no matter what the interpretation of the underlying terms is. For example, suppose you take as your points all of the points on a globe, and lines are now what are called **great circles**, that is, circles on the globe formed by cutting the globe with a flat plane that goes through the center of the globe (see Figure 8.5). Then you find out the first four of Euclid's axioms still hold, but the fifth one, Playfair's axiom, doesn't—any two great circles meet, so there are no parallel lines at all in this geometry. In fact, things are even stranger—in any triangles you draw, the sum of the angles is always *more than* 180°!

In any event, "nonstandard," unexpected models for common ones can be a source of humor as well, just from the surprise factor. There used to be jokes going around like "What is black, white and red all over?" Answers ranged from "A newspaper" (a play on the word "red" versus "read"), "a skunk with diaper rash" or "a sunburned zebra." All are different models for the same sentence.

Figure 8.5: Non-Euclidean geometry on a sphere, with lines being great circles.

Life Lesson: Paradoxes and ambiguity, while often unpleasant in real life, are fodder for great jokes!

Exercises

Exercise 8.1.1. Consider the following:

Suppose that $x = y$. Then $x^2 = y^2$, so $x^2 - y^2 = 0$. It follows that $(x - y)(x + y) = 0$, so by dividing by $x - y$, we get $x + y = 0$, that is, $x = -y$. Adding $x = y$ and $x = -y$, we get $2x = 0$, so $x = -0$ and $y = 0$.

What is the problem, if any, with the argument?

Exercise 8.1.2. Describe 100 in a phrase with at most 140 characters.

Exercise 8.1.3. Does the paradox described on page 397 go away if we allow 100 000 characters instead of just 140?

Exercise 8.1.4. In the hotel paradox, how many new people can we accommodate with their own rooms at a time? What if infinitely many people arrive, demanding their own room—as the night clerk, can you help them?

Exercise 8.1.5. In the hotel paradox, suppose the person in room number n agrees to pay you $1/2^n$ many ounces of gold to stay the night, so, for example, the guest in room number 2 pays 1/4 of an ounce of gold (the rooms further away from the front door of the hotel are worth less). How much money will you take in the next morning if no other guests arrive, but every room is full?

Exercise 8.1.6. A series of *Seinfeld* episodes has Jerry and George write a TV pilot for a show about nothing. Why should this be viewed at a meta level?

Exercise 8.1.7. Mel Brooks created a TV series called *Get Smart!* in the 1960s about a secret agent who was extremely inept. Why is this funny?

Exercise 8.1.8. Suppose that we have a sentence that says "There is some x that relates to all other objects."

(a) Find an interpretation consisting of a world of objects and some relation between objects such that the sentence is true in this interpretation.

(b) Find an interpretation consisting of a world of objects and some relation between objects such that the sentence is false in this interpretation.

Exercise 8.1.9. Suppose that we have a sentence that says "For all objects x, there is at least one different object y that x relates to."

(a) Find an interpretation consisting of a world of objects and some relation between objects such that the sentence is true in this interpretation.

(b) Find an interpretation consisting of a world of objects and some relation between objects such that the sentence is false in this interpretation.

Exercise 8.1.10. Complete your own jokes for the question "What is black, white and red all over?"

2   *Limits of Mathematics*

MATHEMATICS, AS GREAT AND AS POWERFUL AS IT IS, HAS ITS LIMITS (PUN INTENDED!). Sometimes the limits are in terms of our mathematical abilities. But sometimes they exist independently, where no amount of additional thought or expertise will help.

Randomness versus Determinism

We think that certain outcomes are random—the toss of a coin, the roll of a die, whether we get struck by space debris. But are any of these truly random, completely unpredictable? Or, if we knew enough, could we predict exactly what would happen? If we knew the side facing up and the angle a coin sat on our thumb before the toss, the force applied to the coin, the air currents in the room, and so on, if we knew <u>all</u>, could we know before the coin landed what side would turn up? Einstein certainly thought so; he was quoted as having said "God doesn't play dice with the world."

And yet we find it impossible to make accurate predictions about many things. Even the weather predictions beyond say two weeks elude even the most brilliant climatologists. Understanding the problem may lie in a field of mathematics called **chaos theory**, where you have processes that are so involved, that have so many iterations, that future outcomes appear random, even though they're not.

Let me give an example. Suppose we take the function $f(x) = 4x - 4x^2$. If we start with $x = 0.1$, we find that

$$f(0.1) = 4(0.1) - 4(0.1)^2 = 0.36,$$

a simple calculation to do by hand or with a calculator. But now let's repeat the function (that is, iterate it) with the output from the last step, 0.36:

$$f(0.1) = 4(0.36) - 4(0.36)^2 = 0.9216.$$

This is a bit more difficult, but certainly something that we can do, and we are certain of the result.

Now, what if we iterate the process a thousand times? When I use my computer, I get 0.560447527. Of course, my computer's program only uses 10 digits for the calculations, but I should trust my computer, right? Certainly the answer, though likely not exactly 0.560447527 due to some rounding that must go on, must be pretty close, right? I reran the calculation, but with 50 digit accuracy (I can set the number of digits to use in the program), and I get a number starting 0.2451..., a markedly different number, not even close. And if I use 100 digits of accuracy, I get 0.01047.... Is this closer? If I bump the number of digits to 200 I get a number starting with 0.83450..., and I'm more confused than ever. Which is closest to the real answer? I have no confidence, based on the numbers, in whether increasing the number of digits I use will improve the accuracy. The

This problem is often referred to as the principle that "Chaos destroys every computer."

numbers are *chaotic*, they bounce around a lot, and are very sensitive to the number of digits of accuracy I choose to use. Even using a better computer with more digits will not significantly help—eventually the problem will rear its ugly head if I iterate enough.

Real-life problems like predicting the weather, the stock market, or whether you will avoid accidents down the road all suffer from this problem. The outcomes depend on the effect of many, many, many processes repeating, and the inherent problems of small errors (which we can usually ignore when we do calculations) add up to seeming unpredictability. Perhaps randomness, in these cases, is our best mathematical approximation to what is going on. On the other hand, perhaps randomness is indeed a fact of life. Chances are we won't find out the answer.

Tomorrow Never Knows

And there are even mathematical limits to what mathematics can prove! Some work pushes right up against these limits. French mathematician Pierre de Fermat (1601–1665) claimed in the margin of a book that he had a proof that for any positive integer $n \geq 3$ there were no positive integer solutions to the equation $a^2 + b^2 = c^n$. Hundreds of years went by until mathematician Andrew Wiles, in 1994, completed a deep, complicated proof.

And nowadays some mathematical proofs seem to be beyond human hands, requiring the assistance of computers. The **Four Color Problem**, which arose in the 1800s, states that for any possible map you can draw, if you want to color the countries so that two that share a common border receive different colors, then four colors are always enough. The problem was finally settled in the affirmative by Appel and Haken in 1976, but only after reducing the problem to over 1900 cases and using computers extensively to handle each case. There is no known "short" proof without computers, and it seems likely that some proofs in the future, of important and interesting conjectures, will require more than a human touch.

But finally, there are some statements in mathematics for which mathematics will never be able to tell us whether they are true or not. For example, there is a notion of different sizes of infinity. We say that two collections of objects (or **sets**) A and B are of the *same size* if there is a way to match the objects in A to the objects in B. For example, we can match the set $A = \{1, 2, 3, 4, \ldots\}$ with set $B = \{2, 4, 6, 8, \ldots\}$ by matching 1 with 2, 2 with 4, 3 with 6, and so on—matching each number n in A with its double $2n$ in B. So we get the surprising result that A and B have the same size, even though B sits within (that is, is a proper subset of) A! You'll find, more surprisingly, that the rational numbers, all of the fractions, have the same size as the whole numbers (set A).

But one can show that the real numbers have a bigger size. The idea (due to mathematician Georg Cantor) is that if the real numbers, say just those between 0 and 1, had the same size as the whole numbers, then we would be able to match them off, that is, we'd be able to list all of the real numbers between 0 and 1:

$$r_1, r_2, r_3, \ldots$$

Each of the numbers r_i are decimals starting with 0 followed by a decimal point. Then we form a new real number s between 0 and 1 such that

This proof is known as a *diagonalization* argument.

- the first decimal place of s is 4 if the first decimal place of the first number in the list, r_1, isn't a 4, and is 5 otherwise,

- the second decimal place of s is 4 if the second decimal place of the second number in the list, r_2, isn't a 4, and is 5 otherwise,

- the third decimal place of s is 4 if the third decimal place of the third number in the list, r_3, isn't a 4, and is 5 otherwise,

and so on. It isn't hard to see that s can't be any of the numbers in the list $r_1, r_2, r_3, \ldots$, a problem, as <u>all</u> of the real numbers between 0 and 1 are supposedly in the list. The only conclusion we can draw is that the list can't be complete, and there are more real numbers than whole numbers.

The question that mathematicians have is whether the size of the real numbers is the next size up from the size of the whole numbers (this is the Continuum Hypothesis, due to Cantor).

And while it is either true or false, it has been proven that on the basis of the axioms we usually take for mathematics, it's unprovable, so we'll never, never know. And that is something we can only shake our heads at and laugh!

Life Lesson: It may be reassuring that there are some things that even mathematics can't answer.

Exercises

Exercise 8.2.1. Why do you think that weathermen can make accurate short-term (say over a week) predictions?

Exercise 8.2.2. Consider the function $g(x) = 4x(1 - x)$.
(a) Explain why g is the same function as $f(x) = 4x - 4x^2$.
(b) If you have access to a computer and can program, try out the same iterations as were done for f starting at 0.1. How do the results compare to those given in this section (for the same number of iterations and the same number of digits used)?

Exercise 8.2.3. Do you think randomness exists, or is the world just too complicated and chaotic?

Exercise 8.2.4. How many colors do you need to color the following map so that countries that share a border get different colors?

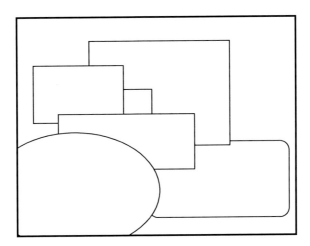

Exercise 8.2.5. Explain why, under the definition of "same size" given in this section, most people have the same number of fingers on each hand and the same number of toes on each foot.

Exercise 8.2.6. Explain why there are as many integers as there are positive integers.

Exercise 8.2.7. (Harder) Explain why there are as many rational numbers as there are positive integers.

And Now the Rest of the Story . . .

In a brilliant breakthrough, Gödel in 1921 proved the impossible—namely, that, provided you took any set of axioms that were consistent, there would always be statements about mathematics that are true, but nonetheless unprovable. His work required him to find a way to encode logical sentences about mathematics as numbers, in what would be known as a **Gödel numbering***. The sentence that he put forward essentially stated "I am unprovable." This sentence is true (as if it were false, it would be provable, and hence true as the axioms are consistent), but unprovable.*

A self-referential sentence, the key to so many previous mathematical paradoxes, was exactly what Gödel needed! He must have chuckled at that. But other mathematicians could only sit back, absorb the result, and appreciate the overwhelming beauty.

As Nazism grew in Austria, Gödel felt more and more uncomfortable. He was not Jewish, but many of his colleagues were, and he found himself out of an academic position. He decided, with his wife Adele, to leave Austria and travel to the U.S., arriving at Princeton University in 1940, where he took up a position at the Institute for Advanced Study. His excellent research continued unabated there.

Also at Princeton was Albert Einstein, and they formed a deep friendship over the years, taking long walks with even longer conversations. Apparently Einstein said that his "own work no longer meant much, that he came to the Institute merely . . . to have the privilege of walking home with Gödel." In December of 1947, Einstein and the economist Oskar Morgenstern went with Gödel to his U.S. citizenship exam. In a funny turn of events, just before his hearing, Gödel told Einstein and Morgenstern that, upon examining the U.S. constitution, he could prove, based on it, that the U.S. could in fact become a dictatorship! The two friends were very nervous as to what Gödel might say to this to the judge, who might treat it as an affront, and might dismiss Gödel without giving him his citizenship. Like out of a sitcom, the judge, who happened to ask Gödel whether the U.S. could become a dictatorship and started to get an analysis, had a sense of what was happening and mercifully cut Gödel off. Gödel got his U.S. citizenship.

Unfortunately, Gödel suffered from mental illness in later years. Fearing that he could be poisoned, he only ate food his wife made him. But when

*Adele was in the hospital for six months in 1977, Gödel refused to eat,
and died, weighing only 65 pounds. A tragic end to a gifted life, one that
continues to resonate inside and outside the mathematical community.*

Review Exercises

Exercise 8.3.1. There is a humorous country song called *Put Another Log on the Fire.* Look up the lyrics online and explain why they are to be interpreted at the meta-level.

Exercise 8.3.2. Suppose that we have a sentence that says "For all objects x, there are exactly two other objects that x relates to."

(a) Find an interpretation consisting of a world of objects and some relation between objects such that the sentence is true in this interpretation.

(b) Find an interpretation consisting of a world of objects and some relation between objects such that the sentence is false in this interpretation.

Exercise 8.3.3. What is the fewest number of colors you need to color the following map so that countries that share a border get different colors?

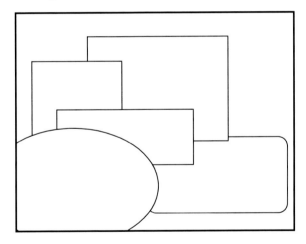

Exercise 8.3.4. Suppose you form the set S of all sets that aren't members of themselves. Why does this lead to a contradiction?

Bibliography

R. Ebert. *A Hard Day's Night (movie review)*. 1999.

M. Lewisohn. *The Complete Beatles Recording Sessions*. Doubleday, Toronto, 1988.

S. Roberts. *King of Infinite Space*. Anansi, Toronto, 2006.

H. Kubo Y. Hagino and G. Sato. *The Beatles Complete Scores*. Hal Leonard, Milwaukee, 1993.

Index